EXPERT
HYPNOSIS
Scripts

For the Professional Hypnotherapist

DR. RICHARD K. NONGARD

Expert Hypnosis Scripts
For the Professional Hypnotherapist
Dr. Richard K. Nongard

Edited by Katie Sanlin

First Printing: December 2016

ISBN 978-1-365-88691-1

PeachTree Professional Education, Inc.
7107 S. Yale, Ste 370
Tulsa, OK 74136
(918) 236-6116
www.SubliminalScience.com

Contents

About the Author

Dr. Richard Nongard is a licensed marriage and family therapist with a Master's Degree in Counseling from Liberty University and a Doctorate in Transformational Leadership (Cultural Transformation) from Bakke Graduate University.

He has written a number of different text-books for mental health and hypnotherapy professionals. His most recent book, *Contextual Psychology: Integrating Mindfulness-Based Approaches Into Effective Therapy* has been a best seller among healthcare leaders.

He is a coach, consultant, and lecturer, offering services to business groups, and seeing clients for clinical hypnotherapy. In the past, his work has included both inpatient and outpatient psychiatric and substance abuse settings, where he's worked with a wide variety of clients over the years.

He has been the president of the ICBCH (International Certification Board of Clinical Hypnotherapists) since 2006 and the ICBCH training program is a great resource for those wanting to master the art of professional hypnotherapy.

Contact Dr. Richard Nongard at www.SubliminalScience.com or (918) 236-6116

I have created a free set of learning resources for readers of this book!

To get your free training videos, additionals scripts and access to our online learning center

Visit this webpage: www.subliminalscience.com/ehs

Skill-Building Contextual Induction Script

Note to the hypnotist

This is a contextual, skill-building hypnotic induction. It is recommended that you record this script and listen to it every day, for a period of ten days. By doing so, you'll discover two things. First, you'll see how easy it is to experience self-hypnosis, allowing yourself to drift into a deep state of hypnotic trance. We can learn from the experience of doing exactly what we ask our clients to do. Secondly, you'll learn how to conduct a hypnosis session. You'll learn patter and scripts, as well as the ideas behind a contextual, skill-building hypnotic induction. You'll also be committing this information to your subconscious mind, so you'll find that when you're doing a hypnosis session with a client, it will be easy for you to guide them to the same experiences that you've been able to create for yourself. This will give you the confidence to do a perfect hypnotic induction with those you are working with.

Induction

For this exercise, find a comfortable chair where you can sit. Or, if you'd like, you can also lie on the floor or the bed. You might also find that it's beneficial to hang a "do not disturb" sign on the door. Turn off your cellphone and turn down the volume on any social media or computer that may be in the room with you.

Once you've found a comfortable place where you can learn, listen, and relax, just let your arms simply rest next to you, or on your lap. Uncross your legs and scan the body for any place you might be carrying the obvious tension of the day. Simply let go of that tension and let those muscles relax. If at any time during this session you need to adjust for comfort, swallow, or even scratch an itch, that's perfectly okay. Those things won't disturb you. In fact, doing those things will simply help you to become even more comfortable and to enjoy your session even further. If there's any noise from outside of this room or even from inside of this room, like a car pulling up, a plane flying overhead, somebody knocking on the door, or a phone ringing, those things won't demand your attention either. In fact, you'll experience them as they are, simply as sounds of the world around you, in this moment. They actually help you to recognize that you're in exactly the right place, doing exactly the right thing, and learning exactly what's most important to you.

Begin this session by bringing your attention to the far wall of the room. There's a spot on that wall. Perhaps it's a shadow, a fleck of paint, or the edge of a picture. It may even just be a blemish on the wall. Wherever you choose, just bring all of your attention to that spot. Just stare at the point you have selected for a moment.

Even if something else in the room should capture your attention, continue to remain fixated on that spot. Focus all of your attention there for a couple more moments. This is really the first point of this exercise because it demonstrates that no matter what else we're experiencing, seeing, or feeling, we can choose where to place our attention. In this case, it is a spot on the far wall.

Any time that you stare at a spot for a long enough period of time, the eyes become a little bit tired. You may have even heard a Hollywood hypnotist say, "Your eyes are getting sleepy." That's not because hypnosis produces sleep, but because when we fixate our attention in any one place for a long enough period of time, the eyes simply become tired. By now you may have noticed that spot has changed a bit. Perhaps it has become crisper, sharper, and clearer, as everything else fades into the background. Or perhaps you've found it's become a little bit hazier, fuzzier, and even more difficult to see. Either way is fine. Perhaps it's simply stayed the same as when you began to focus your attention, and that's okay as well. The one thing you certainly will notice is that if you close your eyes down now, it feels really good. Just simply let those eyes shut completely.

You also notice that even though the eyes are closed, you can still focus your attention at that spot on the far wall. It's almost as if you have x-ray vision. It's really remarkable how we can choose where to bring our attention or our focus. In fact, you can even bring your awareness from that spot over on the far wall, all the way into your mind's eye to that part of the mind where intuition is, where creativity lives, and where learning takes place. Shift your awareness from that point on the far wall, as if you're moving it inside of your mind's eye and into that part of the mind that's creative and enjoys learning and experiencing new things. That's really what hypnosis is. It is something that we learn, something we can experience, and something that can benefit us in many different ways.

Again, focus on the body and anywhere you're carrying any obvious tension of the day. Let that tension go. You can even relax the tiny muscles of the brow and the eyelids. You can unclench the jaw as you relax the muscles of the face, or let relaxation extend through the back of the head, the neck, and the shoulders. You can even let your shoulders drop a bit, as you relax. If you're sitting in a chair, you'll probably find that by letting the chin fall a bit towards the chest, it helps you relax even further, extending your sensation of physical relaxation across the shoulders and upper arms.

As you continue to breathe in and breathe out, we're going to focus on creating a state of physical calmness and relaxation. In fact, we're really going to learn the difference between tension and relaxation. As the muscles of your back, shoulders, and arms relax, just extend that relaxation into the forearms, the wrists, and into your hands. You can even extend that sense of relaxation into the tiny muscles of the fingers. Notice how good it feels to let even those little muscles of the fingers relax.

Take your hands, and while you continue to let your body relax, fold those hands into a fist. Just fold the fingers into the palm of your hand and make a fist with both of your hands, while noticing the feeling of tension as you do that. I don't want you to hold your hands in such tight fists that you feel pain, but I want you to really notice the sensation of muscular tension in the fingers, the palms of the hands, the back of the hand, and the wrist, as you hold those fists tightly. Hold that tension for just a moment, noting what that tension feels like. Now, just relax. Slowly open the fingers a little bit. Relax the muscles of the fingers. Extend those fingers completely as they rest and notice that tingly sensation of relaxation. Noting the difference between tension and relaxation is pretty remarkable, isn't it?

Go ahead and tense those hands up again. Make a fist again with both your hands and hold those fingers into the palm, pressing those fingers into the palm and creating a state of tension in those hands. Hold that tension and notice what tension feels like. Relax again by opening those fingers. Open them all the way. Let them rest and notice that as you relaxed the muscles, the feeling of relaxation doubled.

Just note the difference between tension and relaxation, as you continue to relax, by relaxing the muscles of the back and the muscles of the belly. As you breathe in and out, relax the muscles of the buttocks and thighs. Breathe in a state of calm and exhale relaxation. Let the muscles of the calves and the shins relax and extend that sense of physical relaxation into the little muscles of the feet. You're doing perfect, by just letting go.

In this state of relaxation, continue to pay attention to your breath. In fact, you might even notice that while your body has relaxed, your mind continues to wander, think, and drift. That's okay. In a moment we'll get to a point where we relax the mind as well. However, right now, pay attention to this moment. Become an observer of the breath. You've been breathing since the first day of life. We'll continue to breathe until the last day of life and we do it often, without ever focusing on the breath. So, note what it feels like to breathe the air in and follow the breath as it enters the lungs and turns around deep in the lungs, becoming an exhale. The breath really is amazing, bringing oxygen to the lungs and to every cell of the body.

As you continue to breathe in and out throughout the rest of this session, if you note any sensations, feelings, or thoughts, rather than following them, simply use that as an indicator that it's time to bring your attention back to the breath and to be mindful. As you continue to breathe in and breathe out, bring your attention back to your hands. As your hands rest, notice the heavy sensation that relaxation creates and say to yourself, "My hands are heavy. My hands are heavy. My hands are heavy." Let those hands become heavy and relaxed.

Now, think of the word "warmth." It can be the warmth like that which might come from the sun or warmth like that which might come from inside of the body. Say to yourself, "My hands are warm. My hands are warm. My hands are warm." Let your hands become both warm and heavy, noting a sensation of warmth in those hands. You can even say to yourself, "My hands are warm and heavy. My hands are warm and heavy." Notice how easy it is to create a sense of both warmth and heaviness, recognizing that if you can create warmth and heaviness in the hands or the feet, you can really create any sensation in the body, the mind, or in the spirit.

As you now observe the breath, notice that the breath has become smooth and rhythmic and that without any effort, the heart rate has become calm and regular. Bring your attention to the feet. Say to yourself, "My feet are warm and heavy. My feet are warm and heavy. My feet are warm and heavy." Let those feet become warm and heavy, just as your hands have become warm and heavy, and allow yourself to relax even deeper into a state of hypnosis.

Using the creative part of the mind, imagine that you're in a wonderful place. Perhaps it is a place that you've been to before, a place that you imagine going to someday, or even a mystical place of your own creation. Imagine you're outside on a perfect day in this place, underneath a clear, blue sky. As you gaze up into the sky, notice a large, white, puffy cloud as it lazily and leisurely begins to move across the horizon. With the creative part of your mind, simply follow that puffy, white cloud as it moves off into the distance, becoming smaller and smaller. You'll notice that as it becomes smaller and smaller and as it moves off into the distance, it becomes even easier to set aside any distractions from the past, regrets from previous experiences or fears of the future, and to simply focus on this moment.

Let go absolutely and let that cloud carry off any remaining stress and tension, as it eventually disappears, off into the horizon. That point when you realize that cloud has disappeared is that point that we call the resource state of hypnosis. You've done a great job in this induction, by learning and practicing the strategies of effective hypnosis. So, continue to enjoy this experience that you've created. Not only relaxing, but also learning and recharging, while allowing your higher self to experience a sense of satisfaction and confidence. Knowing that what you've practiced today can benefit you for a lifetime, as well as those who you share these techniques with.

In a moment, I'm going to count from one to three. When I do, allow yourself to become energized and aware. Become aware of the room surrounding you and the surface below you. Become aware of the experience that you've created and how exciting it is to not only learn something new, but to experience something new as well.

And so, with the next breath . . . One. Let oxygen fill your lungs and let that oxygen travel through the blood vessels to every cell of the body, bringing a sense of energy and awareness. Two . . . committing now to a daily practice of self-hypnosis, benefiting not only the clients whom you work with, but yourself. You are now feeling energized, stretching out any muscles that need to be stretched. In a moment, when I count to three, just open the eyes, feeling fantastic and ready for the rest of the day. Three . . . opening the eyes, feeling fantastic and ready for the rest of today.

Dr. Flower's Induction

Note: This is an adaptation of an induction that has been around for more than 50 years. Nobody seems to know who the original Dr. Flowers was. But it is a brilliant induction, using numbers, fractionation and suggestion.

Pick a point on the far wall where you can focus your attention. Now, bring all of your attention to that spot. It is easy to enter a state of hypnosis, and a way to do that is to induce eye fatigue. Many people who go through this simple process find that going into deep trance, quickly, is the natural result. Deep trance will simply feel like a state of relaxation, in mind and in body. You will still be able to hear my voice. Hypnosis is not a state of unconsciousness, but rather a state of dreamy relaxation where the mind is open to new experiences. Are you ready? (Wait to receive affirmative answer.)

I am going to count backwards from twenty to one. With each number, close your eyes. Between numbers, open your eyes. For example: Twenty (close eyes) -open-, nineteen (close eyes) –open-, eighteen (close eyes) –open- Perfect. Let's begin.

Twenty . . .
Nineteen . . .
Eighteen . . .

Notice how you have become absorbed in this process, pushing aside any distractions.

Seventeen . . .
Sixteen . . . .
Fifteen . . .

Notice the sense of relaxation and how it becomes harder and harder to open the eyes.

Fourteen . . .
Thirteen . . . .
Twelve . . .

Any time it becomes preferable to keep the eyes closed, just keep them closed. Simply imagine that you are opening them.

Eleven . . .
Ten . . .
Nine . . .

Perfect. Although you can hear my voice, you are very relaxed. You are not asleep, only deeply relaxed.

Eight . . .
Seven . . .
Six . . .

Let go of any stress completely and access the part of the mind where creativity, intuition, and thought is formed. Let any remaining stress disappear from your body.

Five . . .
Four . . .
Three . . .
Two . . .
One . . .

Completely relaxed in both mind and body . . .

Eye Fixation and a Brief Hypnotic Experience – Induction Strategy

Take in a deep breath. Just breathe in and breathe out. As you breathe out, allow a sense of relaxation to become a part of your experience. As you relax, focus on the far wall. There are some spots over there on the far wall. They may be shadows or even dots that are actually in the design of that far wall. Just pick one of those places and bring all of your attention to that spot. As you breathe in and out, just stare at that spot you have chosen. Keep your eyes fixed on that point.

You may have heard a hypnotist in a television show or a movie say, "Your eyes are getting sleepy. Your eyes are getting sleepy." That isn't because hypnosis creates sleep. It is that when you stare at a point on the far wall, you begin to experience what is called ocular fatigue. You'll notice that as you stare at that spot, it becomes harder to keep your eyelids open. It is almost as if your eyes are getting sleepy. You'll also notice that the spot begins to change a little bit. It might become hazier, fuzzier, and more difficult to see. It might even disappear altogether, or you might notice that spot on the far wall seems to become sharper, crisper, and brighter. It might become the only thing you can really bring your attention to, or you might notice that it simply stays the same, yet you would probably be a lot more comfortable letting the eyes close down.

So, just close the eyes down now. When you close the eyes down, you will notice something amazing. That is that even though the eyes are closed, it's almost as if you have x-ray vision and you can still focus your attention on that spot on the far wall. It is pretty cool, isn't it? As you breathe in and breathe out, bring your awareness from that spot on the far wall. Bring your attention to that place on the inside of your eyelids, where you are seeing that spot. This is the part of the mind where awareness is created, where this moment resides, and where hypnotic intuition and learning takes place.

As you breathe in and breathe out, relax any muscles that need to be relaxed. Unclench the jaw. Let the muscles of the brow relax. You can even let your shoulders drop a little bit, as you let go of any tension that remains. You can let your chin fall towards your chest a little bit. As the chin falls towards the chest, you'll notice the muscles in the neck and shoulders can relax even more, sending that relaxation through the arms, the forearms, and even into the hands and the tiny muscles of the fingers. It feels pretty good to relax and use the resource state of hypnosis that you're creating right now, doesn't it?

With each breath and each number that I count, just double the sensation of relaxation. Five . . . four . . . you're doing perfect. Just go all the way down now. Three, two, one, zero . . .

This is that resource state that we call hypnosis. You are not asleep, just deeply relaxed. In fact, during your time in this resource state, you might be paying attention to each and every word that I use, or you might not be paying attention to the words anymore and instead are only hearing every third, fifth, or tenth word. Either way is fine. It doesn't matter if you're in a light level of trance, or if you're in a deep level of trance. What's most important is that you have committed to learning something new and are using this resource as a key to success in life. So, you can continue to apply what you have learned in hypnosis to every aspect of your success in life. Because this is just a demonstration of hypnosis, in a moment I'm going to ask you to open the eyes, sit up, and feel refreshed and ready for the rest of the day.

Awakening

Use the awakening of your choice.

Jet Lag Hypnosis Script

Note to the hypnotist

This session should be recorded for your own use or for the use of your client.

Pre-talk

This session is designed to help you both avoid and overcome the feeling of jetlag. You can use this session before you travel, on an airplane while traveling, and after you arrive at your destination.

Many people ask me how hypnosis can be useful for avoiding and overcoming jetlag. First, hypnosis is a way of learning how to take control over your body, perceptions of fatigue and alertness, and to regulate your sleep. In fact, hypnosis comes from the Greek word "hypnos" that literally means sleep. Secondly, hypnosis for avoiding and overcoming jetlag is a great strategy for changing your behavior, and we know certain behaviors can contribute to your success in overcoming jetlag. These ideas come from pilots, flight attendants, and seasoned travelers who are experienced in crossing multiple time zones in a short period of time. Thirdly, hypnosis for jetlag works because time is an artificial construct and our perceptions about time can be altered through hypnosis.

There are two main parts of this session. The first part is called the hypnotic induction. By following along, you will be learning how to take physical control of your body and relax, even learning how to put yourself to sleep when your mind does not feel ready. It also focuses on teaching mindfulness, a valuable idea since the only time that really matters is now.

The second part of this session will be the suggestions. There are indirect suggestions that will help you to create an ability to distort time perceptions. This is a really valuable skill when you suddenly cross three, six, or even twelve hours out of a day because of jet travel. There are also direct suggestions that you will internalize that promote behaviors and strategies we know can help you recover faster, feel alert when necessary, and rest comfortably when it is time for sleep in a new time zone.

I think you will find this script of particular value when on your journey, using it to make changes while on your flight. But again, it can be used both before a trip to avoid jetlag and after your arrival to promote wellness. At the conclusion of this session, you will be given a choice. You will have the choice to emerge from the session, energetic, awake, and ready for the rest of the day, or the choice to allow yourself to use this time to enter a deep sleep and rest comfortably. Obviously in an office setting, you will choose to come to full alertness. However, as you listen to the recording at your leisure, you can easily choose to drift into a deep state of sleep instead.

Induction

Use your favorite induction.

Deepener

Use your favorite deepener.

Indirect Suggestions

In the creative part of your mind, imagine that you are looking at a large clock with two hands. Focus on it clearly and imagine that both hands are at the 12 o'clock position. It does not matter if this is 12 noon, or 12 midnight. After all, a clock is just a clock. In fact, we know that time is, to a large extent, a construct of our own imagination. We know this because we have the ability to spring forward for daylight savings time or fall back at the end of the year. It is interesting to note that some places do not use daylight saving time, choosing to opt out of changing the clocks all together. Have you ever thought about how it is possible that Arizona, a small part of Indiana, or entire nation, just arbitrarily opts out of changing the clock? Well, the answer is simple. Time is arbitrary, and it only means whatever significance we have attached to it. It is even entirely possible to do away with time zones. A little known fact is that the countries above and below China span four time zones, yet China has decided that no matter how long it extends east to west, it will only have one time zone.

So, as you relax . . . five, four, three, two, one . . . think of that clock with the hands at noon and imagine seeing that clock and realizing that it is no longer important what the numbers on the clock say. In fact, the hands of the clock could point to 6 o'clock, 3 o'clock, or 12 o'clock, and it really wouldn't make any difference, would it?

This is exactly what Pan Am airlines discovered in the 1950's. Up to that point, jet lag did not really exist. However, as new jets were built that were capable of going further and further, east to west and west to east, pilots, flight attendants, and passengers soon discovered the symptoms of fatigue in the daytime or waking in the middle of the new nighttime, as they crossed multiple time zones rapidly. The founder of Pan Am, looking for a solution, reached out to the Rolex watch makers of Switzerland, asking if they could develop a time piece that could alleviate jetlag. The result was the GMT master, which is one of Rolex's bestselling and time pieces. One hand is set to Greenwich Mean Time and the other hand is set to local time. By turning the bezel, a third time zone can actually be tracked. This essentially lets the mind see time exactly as it truly is; a guide for living and something that is fluid, flexible, and even movable. Now you don't need to buy a watch like this to experience the benefits of the concept. It really is the same as visualizing any clock and realizing that time can move and time can change.

It does not matter if you are from the west heading east, from the east heading west, or even if you go from north to south or south to north. You will find that unlike a compass which is fixed like a magnet, time and the meaning we attach to time is always flexible. This means you can let yourself go to bed early, choose to stay up late, rise in the early hours, or sleep a bit longer than most, listening to your body's needs as it adjusts and making choices based on your own internal clock. The same idea will work for a short distance like a transcontinental flight, or a longer transpacific or trans-Atlantic flight.

So, as you continue to relax, let go of any remaining tension. Let it melt from your body, enjoying this moment, whether it be on an airplane, a hotel, or even your home or office. Five, four, three, two, one . . .

Direct Suggestions

We also know that as you prepare for your arrival, be it in a couple of days, later today, or as you begin your journey now that you have arrived, there are certain things you can do to help you adjust comfortably from a physical perspective. Even simple things like being hydrated and drinking water are helpful. A lot of people do not know that airliners are not kept at an ideal humidity, and so it is easy to become dehydrated in just a short period of time. The answer is natural water before, during, and after a flight. So, even if ordering extra water or bringing a large bottle of water with you is not something you would usually do, at this time you will listen to your body and make the choice to consume extra water.

Of course, we know that alcohol can contribute to dehydration, so if you choose to imbibe, moderation is the choice you will make, opting for water with each and every drink. What is amazing is that these choices will come naturally to you, and will be intuitive for you as you travel.

As you continue to relax, know that movement, activity, and exercise all play a role in how our bodies self-regulate. Simple things like stretching can help you feel better. In fact, either now or when this session is over, you can roll your neck if it is comfortable for you to do so. Just stretch it and notice how good it feels to stretch the legs for comfort or any other part of the body that will benefit.

Lastly, pay attention to daylight. Seek daylight and the bright sun when taking a westward flight and allow yourself the luxury of a darkened room after an eastward flight. In fact, you can close up any curtains before going to bed on an eastward flight, choosing to avoid bright light in the morning, but walking in the afternoon when it is safe and comfortable outdoors. Natural light is important. Seek the sun, seek the outdoors, and let nature be your guide in resetting the biological clocks which we know are able to change and adapt.

As you continue to relax . . . five, four, three, two one . . . continue to enjoy this resource state we call hypnosis, knowing that by paying attention to the most important time, which is right now and this moment, you are preparing your mind and body for optimum performance. By acting intuitively on these suggestions, you know that being your very best is quite possible, even after a long journey.

I will give you a few more moments to enjoy this resource state. In a moment, you may either allow yourself to drift into a deep and comfortable sleep, or choose to reorient to the room around you and emerge refreshed from this short break where you have learned hypnosis.

If this is the end of the day or if sleeping is a wise way to pass the time on an airplane, allow yourself to just drift deeper into relaxation and into deep sleep, ignoring the message I give to emerge, awake and refreshed. The good news is that when you listen to this session again, you may choose the alternate ending, depending on what is most beneficial to you.

And so, if it is time to become alert, oriented, and awake, begin by stretching any muscles that need to be stretched. Breathe in and paying attention to the breath.

Awakening

Use the awakener of your choice.

Mindfulness Exercise

Message to the hypnotist

Practicing stillness and mindfulness on a daily basis can assist you in staying present in the moment, beginning to create the ability to change physical experiences and emotional responses from within, and in accessing and utilizing the creative part of the mind. With intention, set aside some time and find a place where you are undisturbed. Turn off the telephone and turn off the computer in order to really dedicate to this daily practice.

Mindfulness Experience

Begin this practice by taking a minute to simply be still and to experience the natural trance state of learning, of quiet, and of setting aside time for yourself. The idea during this minute of silence is for you to focus on your breathing . . . breathing in and breathing out. Don't try to speed up or slow down the breath. Just simply pay attention to it.

Of course, you notice your mind thinking. This is, after all, what minds do. So many people who are learning self-hypnosis seem disturbed by the fact that the mind is doing what it's supposed to do when they are having various thoughts. It should be of no concern to you that the mind continues to think during this moment of silence. The idea here is to not follow those thoughts. Simply acknowledge those thoughts as thoughts and use it as an indicator or a cue to return your attention back to the breath.

Go ahead now and close the eyes. Just release any tension of the day that you might be carrying. Unclench the jaw and let your shoulders relax. Breathe in and breathe out, focusing on the breath. Now you're doing exactly what you're supposed to be doing by setting aside a minute to simply practice stillness, to practice paying attention to the moment, and using that inquisitive part of the mind to observe your experience. With the next breath, let that breath fill your lungs. Let it energize your body for continued learning. Take in another breath, noticing that breath as you inhale and exhale. Now I'm going to count from one to three. As I do, simply become more energized, more aware, and ready for the rest of your day. One . . . two . . . Open the eyes, feeling fantastic.

SELF-HYPNOSIS

Note to the hypnotist

So often, we're busy throughout each and every day and never really set aside time just for the quietness of the mind that allows us to become aware of the various trances that demonstrate that we're experiencing the resources within us. By practicing self-hypnosis on a regular basis, you will learn how to gain access to those states. It is recommended that this script be recorded for your use.

Self-Hypnosis Experience

To begin, simply breathe in and breathe out. Go ahead and close the eyes now and pay attention to anywhere you're carrying the tension of the day. Unclench the jaw and relax the little, tiny muscles in the brow and the eyelids. Rest your hands on your legs and pay attention to your breath. By paying attention to the breath, we're drawing our attention to this moment. You're doing absolutely perfect. You're doing exactly what you're supposed to be doing right now to learn the art of trance utilization.

Anywhere you notice any remaining tension of the day, just release those places that hold that tension. Let your shoulders relax and any stress and tension that remains simply melt away. Notice, without effort, your breathing has already become more smooth and rhythmic. Your heart rate has slowed and become calm and regular. It's amazing how practicing something this simple, with intention, can begin to focus our attention and allow us to access that place of inner creativity we all possess. You're doing perfect.

Now, open your eyes for a moment. Open your eyes. Open them wide. See the light in the room around you. Take in a breath . . . breathe in, breathe out, and notice something. Notice there's a big difference between your state of awareness, the resources you have, the feelings that you have with the eyes open and being fully alert, versus what you were experiencing just a moment ago as you allowed your muscles to relax and allowed any stress or tension to simply disappear.

It's important in self-hypnosis to begin to really pay attention to the difference between tension and relaxation, distraction and focused attention, and between an alert trance and relaxed trance. Now go ahead and close your eyes again. With each breath, double the sensation of relaxation. What you're doing here is practicing the process of bringing yourself to a resource state of relaxation, openness, and creativity. You're doing wonderful. By accessing this creative part of the mind, we have the ability to create new experiences. We not only have the ability to create emotional or mental experiences, but even physical experiences.

So, as you focus on your hands, think of the word "warmth." You might think about warmth like that which comes from the sun or warmth like that which comes from inside of the body. As you think of warmth, focus on your hands and say to yourself, "My hands are warm. My hands are warm." Let yourself notice this sensation of warmth on the back of the hands, on the inside of the hand, and even the fingertips. It might be a slight awareness of warmth, or it may be an incredible feeling of warmth. Either way is okay. You're using that creative part of the mind to create a feeling of warmth in the hands as you relax.

As easily as you can create warmth in the hands, you can even create heaviness as well. Say to yourself, "My hands are heavy. My hands are heavy." Let those hands become heavy. Notice that as you breathe in and breathe out, as the heart rate slows a bit, and as the breath continues to be smooth and rhythmic, your entire body can experience a sense of heaviness. Relaxing deeply, you can actually congratulate yourself. You can congratulate yourself because as you notice your awareness now, you can notice that you've not only created relaxation, but you've accessed that part of the mind where your creativity lies, where intuition exists, and where new solutions can be experienced.

You've done a great thing so far, by simply taking some time to yourself to be still and open to new lessons. You're even going to let a smile come across your face as you feel good about the place that you've created right here and right now. As you continue to breathe in and breathe out, with each number and each breath, double that sensation of relaxation. Five, four, three, two, one. Perfect. Go all the way down now, completely relaxed.

I'm going to give you a few moments here to simply observe this state. Observe this state with curiosity and wonder about the potential that your mind has for creating new experiences, new states, and being aware of new resources. In fact, over the next few moments, allow yourself to associate into a state of learning and into a state of feeling good about the decision you've made to master the art of self-hypnosis. Be open to the possibilities of what you can achieve through a dedicated and daily practice of self-hypnosis.

With each breath you take, relax even further and allow yourself to be comfortable, safe, and warm. With the next breath, simply let that breath fill your lungs with oxygen. Let the oxygen go through the blood cells of the body, oxygenating each and every cell in the body, feeling energized and fantastic from that breath. Breathe in another breath and breathe it in deeply, allowing yourself to feel energetic and excited about the potential for learning new things.

As I count from one to three, be ready to open your eyes in a moment, feeling fantastic and ready for the rest of the day. One . . . pay attention to the breath. Two . . . stretch out any muscles that need to be stretched. And 3 . . . open the eyes, feeling fantastic, and ready for the rest of the day.

Creating Positive Resources States

Note to the hypnotist

Here is a strategy for self-hypnosis that is actually very simple and very powerful to teach clients or as an exercise for yourself. It might be helpful to record it for personal use. For this exercise, you're going to need two things. First, you're going to need a pen and you're going to need a piece of paper. It really doesn't matter what the paper is. You can use index cards if you want to, or you could even use yellow sticky notes. What I do is that I write down the resource state that I would like to create and then I tape it right on my computer monitor. This is really a form of meditation and a way to create the positive emotions which are resource states for you. So, think for a minute about what would be valuable for you to create. Would it be valuable for you to create joy? Would it be valuable for you to create satisfaction? Would it be valuable for you to create confidence? Would it be valuable for you to create a state of calm?

So, just take your pen and piece of paper and write that resource state that's valuable to you on that piece of paper. It's important that you actually use a paper and pen because there's really power in taking a thought or an idea and making it concrete and real by writing it out. This is really the same power as the spoken word and we know that from a metaphysical perspective, this spoken word has all kinds of power to create. In Genesis Chapter 1, everything that was created was created when God spoke. We understand, throughout history, the power of the spoken word. The American metaphysicians of the 1800s and 1900s, people like Charles Bell, understood the power of the spoken word. When we speak, we create. The idea here is really simple. Nothing exists in this world unless it was an idea first. When we take an idea, we make it real by writing or speaking it. Then it becomes something that we can possess, own, claim, or utilize. So, just place your word on your computer monitor or on the desk in front of you.

A lot of people have a misbelief that hypnosis requires the eyes to be closed. That's simply not true. We can do hypnosis with our eyes open at any time. If you were to come to see me for hypnotherapy at my office, at some points I would have you open your eyes and at other points I would ask you to close your eyes. We can achieve trance at any time because we're always in trance. The important part is how we utilize it, and often we can utilize trance with our eyes open.

Suggestive Therapy

Now that you've placed that word on a table or desk in front of you or you stuck it on the front of your computer monitor, simply take in a deep breath. Breathe in, breathe out, and pay attention to the breath. You neither have to speed up or slow down the breath, but begin to fixate your attention on that word which you wrote. That word represents the resource state that would be valuable to you in the arena of emotional intelligence, and that is a resource state that you can amplify or step into. It is a resource state that you have the ability to create.

So as you breathe in and breathe out, focus your attention on that word which you've written. Anywhere in the body where you notice the muscles are tense, relax those muscles. Unclench the jaw and let the shoulders drop a bit, simply keeping your attention focused on that word in front of you. Now, you might notice something. You might notice that the handwriting on that piece of paper becomes a little bit brighter, sharper, crisper, and clearer because you've been staring at it for a period of time. It is almost as if it's jumping off the page. Or, you might notice that word becoming fuzzier and less clear. It might even appear a bit out of focus. Maybe it's even as if the word is becoming a part of that piece of paper. You might not even notice any change at all. Any experience here is okay. There's not a right or a wrong way to experience self-hypnosis.

So as you breathe in and breathe out, focus on the breath. At the same time, be aware of that word. You can continue to keep your eyes open and focused on that piece of paper if you would like to, or you can go ahead and close the eyes down now. Even if the eyes are closed, you can still see the word on that piece of paper, almost as if you have x-ray vision. You've taken this time for self-hypnosis today to learn a new strategy. It is the strategy of emotional intelligence and creating a resource state. So, as you breathe in and breathe out, focus on not only seeing that word, but experiencing that word. Allow yourself to feel a sensation associated with that word.

For example, if the word was confidence, notice it in your posture. If the word was joy, notice a feeling of joy. Simply become aware of the physical experience of the word which you've written. Allow yourself to experience, on a physical level, the reality of that state. As you breathe in and breathe out, relaxing and taking time to learn something new, on an emotional level be aware of that word. You can even say it in your mind. For example say, "I feel joy. I feel joy. I feel joy," or whatever the word is that you've chosen. Go ahead and say that to yourself.

Notice that each time you say that word, it gives power to the word that you've written down. As you breathe in and breathe out, you can even congratulate yourself. You've done a great job with self-hypnosis today. Not only have you chosen this word, but you have allowed yourself to both physically and mentally associate fully into that resource state. It feels pretty good, doesn't it? The great thing is that you don't need this session to be able to do this at any time, any place, or anywhere. In fact, from this point forward, should you need to tap in to that resource you identified, you can simply close your eyes for a moment or focus on an object in front of you for a moment. Take in a breath and bring yourself back to this state which you've created here and now. I'll give you another moment or so to enjoy this resource state that you've created. Just enjoy the feeling, enjoy the awareness, even using positive affirmations as a tool to amplify that experience.

Now it is time to take in a breath and to reorient yourself to the room around you. Just focus on that word in front of you, noting the power in the written word and the spoken word. Take this moment in your busy day to simply focus your attention on an ability that you have within you to create a resource state. Take in another breath. As you take in that breath, just let the air fill your lungs. Let the oxygen flow to every cell of the body. Notice how wonderful it feels to take some time to learn something new and to associate into a wonderful resource state. Take in another breath, feeling wide awake, fully alert, and open your eyes, feeling fantastic.

Self-Hypnosis Script for Confidence

Note to the hypnotist

For self-hypnosis purposes, you can make a recording of this script, or you can simply bring yourself to that place of self-hypnosis and read this to yourself, letting the words come alive and becoming an affirmation to yourself.

Suggestive Therapy

Begin by saying to yourself, "I feel completely relaxed and in control. I feel completely relaxed and in control. I feel completely relaxed and in control." Think back to a time when you were successful at your activities and a time when you reached and exceeded your personal performance expectations. Perhaps everyone cheered for you because you did so well. That was one of the happiest times for you. You were in control and confident in your actions. You knew what to do, how to do it, and you did it well. Say to yourself, "I know what confidence is. I know what confidence is. I know what confidence is." In your mind, recreate the very emotions that you felt at that time. Re-experience the power you felt, the success you achieved, and the excitement of the moment. Allow yourself to feel those feelings again now.

Imagine you now have an imaginary stick of chalk in your hand. With that chalk, draw a big white circle. Visualize yourself drawing that circle. Just seeing this circle instantly creates wonderful feelings for you. This is your circle of confidence and success. Say to yourself, "In my circle, I feel confident. In my circle, I feel confident. In my circle, I feel confident." You created this imagery, and with our thoughts we create reality. You will always have a stick of chalk with you in your mind and you can draw this big white circle around you and your tasks at any time you wish, for the rest of your life.

Step into your circle. Feel the power and control of the circle. From this point forward, when you are inside your circle, you will instantly begin to feel confident and successful, as you have felt before when you performed so well. From this point forward, anytime you begin to feel anxiety, worry, or stress about your abilities, you can simply step into your circle of confidence. Those worries and stressors will simply melt away, leaving you feeling confident, successful, and refreshed. You have confidence in your ability to succeed with every move you make. When you step into that big white chalk circle that surrounds you and your life, you will be able to take consistent action to achieve your personal performance goals. Say to yourself, "I have confidence to succeed. I have confidence to succeed. I have confidence to succeed." Begin to make important changes in our life.

Ultimate Hypnosis Script

Note to the hypnotist

I call this the "Ultimate Hypnosis Session" because the goal is to learn to experience deep states of relaxation and it intentionally addresses each of our five senses. Rather than focus on one technique of visualization, or simple auditory affirmation, it combines these learning styles. This allows you to develop acuity in each of these five areas. Additionally, it is structured in a way that any suggestion you have for yourself, such as those in the areas of health, habits, and personal improvement, will easily be accessed by the subconscious mind. Even multiple goals can be addressed by participants in this session.

The goal for this script is to listen and relax, rather than to think and to take action. Often in self-hypnosis, we overlook the value in using guided self-hypnosis to passively enjoy the hypnotic experience. This script is designed to let you do just that. The easiest way to do this is to make a recording of this script and to find a comfortable and quiet place where you can relax.

Autogenic/Relaxation Induction

Now, take a breath in. Breathe in and breathe out. As you breathe in and breathe out, go ahead and close the eyes down. As you breathe in and breathe out, pay attention to the breath. You don't have to speed up or slow down the breath. Simply pay attention to it. Scan your body and notice anywhere where you're carrying the tension of the day and simply let that tension disappear. Drop the shoulders and unclench the jaw. You can even drop your chin towards your chest a little bit if you'd like. Although this helps you relax, at no time are you going to be asleep. Instead, you will simply be deeply relaxed, learning new skills and strategies.

As you pay attention to your breath, you'll probably notice that your breathing has already become smooth and rhythmic without any effort on your part. The heart rate has probably even slowed a little bit, becoming calm and regular. As you continue to breathe in, relax the muscles of the brow, the muscles of the jaw, and the muscles of the neck and shoulders. Continue to breathe in and breathe out. It feels pretty relaxing, doesn't it? Now, go ahead and open your eyes for a minute. Open your eyes, just for a moment, and check in with yourself. Ask yourself if you are a little more relaxed and calm now than you were a moment ago.

It's pretty amazing how easily we can create a new state so rapidly, by just closing the eyes, letting the tension go, and letting some muscles relax. Go ahead and close the eyes down again, breathing in and breathing out. Bring yourself to that point of relaxation where you were just a moment ago, maybe even noticing yourself doubling that sensation of relaxation with each breath. Notice the arms becoming relaxed and heavy from relaxation, the loosening muscles in the back and the belly, and even the muscles in the buttocks and thighs.

As you breathe in and breathe out, bring your attention for a moment to your hands. Notice how you can even relax the little, tiny muscles of the hands, and even the tiny muscles of the fingers. Just let those hands relax and be completely supported. When they're completely relaxed, you will find that the relaxation causes you to become aware of a sense of heaviness. Think of the word "heavy" for a moment. Think of your hands and say to yourself, "My hands are heavy. My hands are heavy." Really notice how heavy your hands are. They are so heavy that if you tried to lift your hands, you'll notice they become locked down, heavy from relaxation. Even if you tried to move your hands, you can't move your hands because they're so heavy and so deeply relaxed.

So, breathe in again and breathe out, thinking of the word "warmth." Think of the warmth like that which might come from the sun, or warmth like that which might come from inside of the body. As you focus on your heavy hands, continue thinking of the word "warmth." As you focus on your hands, thinking of that word, allow yourself to notice a sensation of warmth in your hands. You can notice both a sense of warmth and heaviness. Say to yourself, "My hands are warm. My hands are warm. My hands are warm." Notice that feeling of warmth that comes from inside or that can be felt on the back of the hands as they are at rest. As your body relaxes, your legs, your thighs, your calves, your shins, and even your feet can relax.

Even the little, tiny muscles of the toes can relax. You can notice a sense of heaviness in your feet and the muscles relaxing.

In fact, think of the word "heavy" again and say to yourself, "My feet are heavy. My feet are heavy." Think of the word "warmth." As you think of that word, notice a sensation of warmth in those feet. Just let those feet be both warm and heavy. Say to yourself, "My feet are warm and heavy. My feet are warm and heavy." It's amazing how we can actually create a sensation of warmth and heaviness, even though we haven't adjusted the thermostat.

As you breathe in and breathe out, notice the heart rate is calm and regular and the breath is smooth and rhythmic. Bring your attention to the forehead, across the brow. As you pay attention to your forehead, think of the word "cool." Allow your forehead to experience a sense of coolness. Say to yourself, "My forehead is cool. My forehead is cool. My forehead is cool." You're doing perfect, having created both an awareness of warmth in one part of the body and coolness in another part of the body, as well as creating a feeling of heaviness in the hands and feet. Five, four, three, two, one.

The lesson here is that you can create any sensation, thought, or experience. What you can achieve here, you can achieve there, and by thinking thoughts such as "relax", "cool", "heavy", or "warm", you were able to notice those things and create those experiences. What else would you like to create? Love? Healing? Change? Calmness? Energy? I do not know what is most important for you to create in this time that you have set aside. Maybe your conscious mind is not even aware of what it is, but a part of you knows what is most important for you to create today. Whatever that thought, feeling, or experience is, be it known or unknown to your conscious mind, it is a part of your subconscious awareness. Continue to relax, breathing in and out, and know that by setting this time aside for your own well-being, you will manifest that which is most important to you today.

Now, notice how you feel physically. Notice the feeling of heaviness in the muscles as they relax. Notice this sense of heaviness in the eyelids and even notice the weight of your resting hands. They feel so deeply relaxed and heavy that even though your mind knows that you could move them, moving them isn't something you care to do. You are enjoying this stillness, if even for a brief moment during a busy day. Now, notice your feet also feeling very heavy and deeply relaxed. Perfect.

With the creative part of the mind, imagine the stressed version of yourself, heavily relaxing. Also, imagine a new you, a courageous you, composed in that part of the mind where creativity exists. This "courageous you" begins to feel a sense of lightness. It is a lightness that transcends any stress and a lightness that feels as if it can rise above you. Feel that creative and courageous you, now. It is that part of you that knows it can handle any situation and that part of you that is capable of healing during difficult times. Let that feeling of lightness begin to float out of you and above you, as if the real you has found a way to rise above the stress. As you both relax and float, allow that creative part of your mind to continue to imagine the real you, the creative and courageous you, rising above all stress. You can even travel high up and far away to see yourself transcending both the place where you are and the time constraints imposed by stress.

Allow yourself to enjoy the feeling of being beyond any limitations, deadlines, or stress. Just feel the sensations of serenity, peace, and freedom. You can let your mind drift, dream, and float, becoming detached from any worldly stress. Rise above to a higher place, a place where your higher self can truly succeed and use all of the internal strengths that are deep within you. Pay attention to your feelings of confidence, freedom, and joy. These are not states that I have created, but rather states that you have created by simply taking a moment for yourself to set aside stress and tap into your inner resources. It feels good to tap into these inner resources and to confidently know that these can be accessed at any time.

Unite that higher self with the old self, just like a soap bubble gently floating to its resting spot. Allow that lighter, higher self to drift back into an awareness of your relaxed and heavy body. You realize that while this mental exercise has been an interesting experience, you always carry the ability to feel and see beyond stress. As I count backwards from five to one, deepen your hypnotic state. Five, four, three . . . let yourself relax completely. You are not asleep, but deeply relaxed . . . two, one, zero.

Over the next couple of minutes, I am going to guide you through a process of dissociating here from there. Milton Erickson called this the "Nowhere Technique." As you relax, you can recognize that a part of you is here. You can feel the surface below you, while you have your eyes closed and your mind relaxed. You can also see that a part of you is drifting away. Drift to a place that is really nowhere. In fact, drift to a place we call the middle of nowhere. You can let your mind drift and meet me in this place, which is a place that has no time. It is a place that has no place, in the middle of nowhere. It is a place that has these words and your awareness. It is a place that is neither here nor there. It is just a place of your own creation.

In the middle of nowhere, there is no awareness of pain, but simply an awareness of nothingness. Here in the middle of nowhere, nothingness is just fine. There is nothing to be and nothing to feel. There are no feelings to feel in the middle of nowhere. It is a very pleasant place to be, isn't it? It is here that you can identify what is most important to you, such as a state of healing, a state of change, or a state of empowerment. It is here that you can release now and forevermore, anything either known or unknown to you that was once keeping you from experiencing success. Take some time now to reward yourself with a few moments of tranquility, having met your needs today in this creative time you have set aside for yourself.

Awakening

Use the awakening of your choice.

Chemotherapy-Induced Nausea and Vomiting (CINV) Script

Evidence

Richardson, J., J. E. Smith, G. McCall, A. Richardson, K. Pilkington and I. Kirsch. "Hypnosis for Nausea and Vomiting in Cancer Chemotherapy: A Systematic Review of the Research Evidence." European Journal of Cancer Care 16, no. 5 (2007): 402-412. To systematically review the research evidence on the effectiveness of hypnosis for cancer chemotherapy-induced nausea and vomiting (CINV). A comprehensive search of major biomedical databases including MEDLINE, EMBASE, ClNAHL, PsycINFO and the Cochrane Library was conducted. Specialist complementary and alternative medicine databases were searched and efforts were made to identify unpublished and ongoing research. Citations were included from the databases' inception to March 2005. Randomized controlled trials (RCTs) were appraised and meta-analysis undertaken. Clinical commentaries were obtained. Six RCTs evaluating the effectiveness of hypnosis in CINV were found. In five of these studies, the participants were children. Studies report positive results including statistically significant reductions in anticipatory and CINV. Meta-analysis revealed a large effect size of hypnotic treatment when compared with treatment as usual, and the effect was at least as large as that of cognitive–behavioral therapy. Meta-analysis has demonstrated that hypnosis could be a clinically valuable intervention for anticipatory and CINV in children with cancer. Further research into the effectiveness, acceptance, and feasibility of hypnosis in CINV, particularly in adults, is suggested. Future studies should assess suggestibility and provide full details of the hypnotic intervention.

Note to the hypnotist

This specific script was not studied as part of the aforementioned meta-analysis. The citation was provided to help you understand that hypnosis is an evidenced-based approach and to give you confidence is using the methods you employ.

Autogenic/Relaxation induction

Now, take a breath in. Breathe in and breathe out. As you breathe in and breathe out, go ahead and close the eyes down. As you breathe in and breathe out, pay attention to the breath. You don't have to speed up or slow down the breath. Simply pay attention to it. Now, scan your body and notice anywhere where you're carrying the tension of the day and simply let that tension disappear. Drop the shoulders and unclench the jaw. You can even drop your chin towards your chest a little bit if you'd like. Although this helps you relax, at no time are you going to be asleep. Instead, you are simply and deeply relaxed, learning new skills and strategies.

As you pay attention to your breath, you'll probably notice that your breathing has already become smooth and rhythmic, without any effort on your part. The heart rate has probably even slowed a little bit, becoming calm and regular. As you continue to breathe in, relax the muscles of the brow, the muscles of the jaw, the muscles of the neck and shoulders, and continue to breathe in and breathe out. It feels pretty relaxing, doesn't it?

Now, go ahead and open your eyes for a minute. Just open your eyes for a moment and check in with yourself and ask if you are a little more relaxed and calm now than you were a moment ago. It's pretty amazing how easily we can create a new state so rapidly, just by closing the eyes, letting the tension go, and letting some muscles relax. Go ahead and close the eyes again, breathing in and breathing out. Bring yourself to that point of relaxation where you were just a moment ago, maybe even noticing yourself doubling that sensation of relaxation with each breath. Notice the arms becoming relaxed and heavy from relaxation and just going ahead and loosening the muscles in the back and the belly, as well as the muscles in the buttocks and thighs.

As you breathe in and breathe out, bring your attention to your hands. Notice how you can even relax the tiny muscles of the hands, and even the little muscles of the fingers. Just let those hands relax and be completely supported. When they're completely relaxed, you will find that relaxation causes you to become aware of a sense of heaviness. Think of the word "heavy" for a moment. Think of your hands and say to yourself, "My hands are heavy. My hands are heavy." Really notice how heavy your hands are. They are so heavy that if you tried to lift your hands, you'll notice they become locked down, heavy from relaxation. Even if you tried to move your hands, you can't move your hands because they're so heavy and so deeply relaxed. You're doing perfect, by the way.

Breathe in again and breathe out, and think of the word "warmth." Think of the warmth like that which might come from the sun or warmth like that which might come from inside of the body. As you focus on your heavy hands, think of the word "warmth." As you focus on your hands, thinking of warmth, allow yourself to notice a sensation of warmth in your hands. You can notice both a sense of warmth and heaviness. Say to yourself, "My hands are warm. My hands are warm. My hands are warm." Notice that feeling of warmth that comes from inside or that can be felt on the back of the hands as they are at rest. As your body relaxes, your legs, your thighs, your calves, your shins, and even your feet can relax.

Even the little, tiny muscles of the toes can relax. You can notice a sense of heaviness in your feet and the muscles relaxing. In fact, think of the word "heavy" again and say to yourself, "My feet are heavy. My feet are heavy." Think of the word "warmth." As you think of the word "warmth," notice a sensation of warmth in those feet. Just let those feet become both warm and heavy. Say to yourself, "My feet are warm and heavy. My feet are warm and heavy." It's amazing how we can actually create a sensation of warmth and heaviness, even though we haven't adjusted the thermostat.

As you breathe in and breathe out, notice the heart rate is calm and regular and the breath is smooth and rhythmic. Bring your attention to the forehead, across the brow. As you pay attention to your forehead, think of the word "cool." Allow your forehead to experience a sense of coolness. Say to yourself, "My forehead is cool. My forehead is cool. My forehead is cool." You're doing perfect, having created both an awareness of warmth in one part of the body and coolness in another part of the body, as well as creating a feeling of heaviness in the hands and feet. Five, four, three, two, one.

Suggestive Therapy

As you relax, you've now entered the resource state that we call hypnosis. It isn't a state of sleep, but a state of deep relaxation. You've created this sense of relaxation here and now, which should give you great hope. That is because if you can create relaxation at any level, especially during a period of life that is stressful to you, then you have a great ability. This is the ability to create, from within yourself, any state which is helpful to you.

This is one of the great things about learning something new like hypnosis. It can be a great benefit to you as you use your newfound skill to change some of the undesired feelings that the treatment process might produce. You might even ask yourself how such powerful feelings or automatic responses like nausea and vomiting can be changed. After all, aren't they automatic responses that can't be controlled? Well, you've already demonstrated, just in the past few minutes, an ability to create simple feelings like warmth or heaviness, even though no one changed the heater setting or added weight to your hands. These are sensations that you became aware of and amplified. They too were automatic responses to suggestion. The lesson here is that you can control what suggestions you wish to respond to and at what time you would like to respond to those suggestions. They could be suggestions of side effects which would be unpleasant or they could be suggestions that you create that can counter negative experiences.

Are you ready to try this and begin to use such a strategy? It is a strategy where you can choose which suggestions to respond to. Great! You can control what your mind sees, can't you? You have an ability to create almost any mental picture, at any time, don't you? Here you can actually test this. As you continue to relax, think of a box. Now, think of a book and then think of a flower. Did you think of a flower? Did you think of a box? Did you think of a book? It's amazing how easy it is to visualize instantly.

Now, think of the clear blue sky and a single, white, puffy cloud that is leisurely and lazily moving off towards the horizon. It will do what clouds do and become smaller and smaller, eventually disappearing off into the horizon. Perfect. You're really good at creating images, aren't you? Have you ever seen a gentle waterfall? Maybe you have seen one in real life or maybe even just in pictures. They always bring a sense of calmness to me. Last year when I was in Hawaii, I was at a beautiful place on the big island where I was actually able to stand underneath the cool and refreshing water of a gentle and beautiful waterfall. It felt refreshing to me. The water was coming down onto my head and shoulders and flowing down into the pool below where I was standing. Can you imagine what that might be like? Maybe you've even experienced something like this too. Even if you haven't experienced it though, I bet you know what it would feel like to feel the cool water coming down, bathing the body with refreshment and relaxation.

Take a moment and actually seeing yourself underneath the gentle waterfall, feeling the feelings that you would feel if you were there. It would probably feel really good, wouldn't it? It might even feel comfortable and healing. Let yourself feel those feelings and when you find yourself really enjoying this sensation, almost as if you feel like you're really underneath that healing cool waterfall, recognize that sense of comfort and gently squeeze the thumb and forefinger on one of your hands together. Go ahead and squeeze those fingers together. By doing this, you're capturing both the image and the feeling of comfort that you've created.

Now, this is a really interesting juxtaposition of the experience of how our food and drink are supposed to travel down like a waterfall through the body, eventually even landing in a pool, much like a waterfall. But in fearing nausea, we begin to create an image that is the opposite of what the body usually does. When you have a thought like an anticipatory fear, or even a physical sensation in the body produced by treatments or medication, just take time out for yourself. Take a deep breath, close the eyes, and press those fingers together. This will bring you back to that mental image that you created just moments ago of being underneath that cool and refreshing waterfall, as that water cascades over the head and shoulders, down the torso and into the pool below.

If you were to have a sensation that's distressing to you, you'll be amazed of your ability to instantly change that state. You can just see the visual imagery that you've created, of the way our food our drink is supposed to travel, just like a cool, calm, and refreshing waterfall or shower. As you continue to relax . . . five, four, three, two, one, zero . . . you are not asleep, but deeply relaxed. Continue to breathe in and breathe out and think back to your hands and the sense of warmth you were able to create or become aware of within them. Say to yourself, "My hands are warm. My hands are warm." Think of the feeling of heaviness you were able to create in your feet and in your hands. Say to yourself, "My hands and feet are heavy. My hands and feet are heavy." Think back to the beginning of this session when you focused on your forehead and said to yourself, "My forehead is cool. My forehead is cool." You felt a refreshing sense of coolness across that forehead.

Now, I don't know which sensation is most valuable to you. Perhaps it's a sense of heaviness and stability as you rest. Perhaps it's a sense of warmth, or perhaps it's a sense of coolness that is most likely to earn you relief from a feeling of nausea. Perhaps there's even another sensation which is the opposite of nausea that is valuable to you. The wonderful thing about the mind and the body is your ability to create that awareness or that sensation which is comforting to you at any time and at any place.

As you breathe in and breathe out, you can say to yourself, "My heart rate is smooth and regular. My heart rate is smooth and regular." Maybe that's the sensation you can become aware of to help you feel stable and well. You can say to yourself, "My breath is smooth and rhythmic. My breath is smooth and rhythmic." Maybe by paying attention to the breath, it is that awareness which will allow you to feel the greatest level of comfort, even though you may have heard a suggestion from another medical professional that a treatment might produce a side effect.

As you continue to relax, take another minute before you open your eyes. Take another minute to just be aware of your body and this sensation. Become aware that you have an ability to create feelings of warmth, heaviness, and coolness in the body, smooth and rhythmic breathing and heart rate, and feelings of confidence, comfort, and relaxation. It's amazing how during this time, which is only a short time of learning, you can truly master a skill of creating from within any resource state, either physical or emotional, that is valuable to you. In fact, you can even test your ability to do that.

Earlier in this session, I asked you to test heaviness by having you lift the hands, but they became so comfortable and relaxed that they became locked down. Of course, you could also think of lightness if it would be valuable to you and easily lift the hand half an inch or an inch from where it is resting. As you continue to breathe in and breathe out, press those fingers together one more time. Just press your thumb and forefinger together. Simply squeezing those fingers together brings you right back to that cool, refreshing feeling of the waterfall and the feeling of the waterfall coming down from above onto your head. It washes the body, falls into the pool below, and is carried away.

That thought you created moments ago is that state that you experience right now. This gives you the ability over the coming days, months, or even years, at any time or place, to choose that sensation which would be most valuable for you to create and pay attention to. By either recalling this experience of using that anchor of pressing your fingers together, or by using your new skill of autogenic training, you can create heaviness, lightness, calmness, coolness, warmth, or comfort.

You've done fantastic in this setting and in this session. In fact, you've done so well and have certainly followed this process in great detail, investing in its outcome. So, I have no doubt that the situations you may have feared or where you may have even previously felt nauseous, you now have an ability to create a new experience that's truly valuable to you.

Now, be ready in a moment to open the eyes, feeling fantastic. Before you do that, pay attention to the surface below you and to the air in the room around you. Reorient yourself to this time and this place. Take in a breath. Feel the air enter the lungs. Pay attention to the air as the oxygen spreads to the blood supply, energizing the muscles of the body. Take in another breath. Become ready, in a moment, to open the eyes and even stretch out the muscles that need to be stretched. One, two, three . . . open the eyes and feel fantastic and ready for the rest of the day.

Overcoming Pain and Increasing Physical Comfort

Pre-talk

Hypnosis is an effective tool for managing pain. The reason is simple. It changes your awareness and teaches a person how to overcome the pain by increasing the level of comfort they feel. This hypnosis session is not a substitute for medical intervention. Pain is a resource of the body that should be listened to, as it can alert us to medical needs that require attention. But for those already under the care of a physician or those enduring chronic pain, learning the skills of hypnosis is an excellent way to see pain from a new perspective. By using hypnosis, you can remove the hurt from the pain and create an awareness that can help you to become more mindful of life outside of pain.

Hypnosis is a natural process. Each person enters hypnosis daily, as we cycle from high levels of alertness to the slower periods of the day, and then into deep sleep. Hypnosis is a form of relaxation coupled with focused concentration. It is a time to use the creative and intuitive parts of the mind to help the body experience something new. It can help us to recover from illness, accident, or trauma, by using the natural resources within us.

All hypnosis is really self- hypnosis and there is not a right way or a wrong way for you to experience this learning process. You don't even have to try to be hypnotized. After all, you already know how to experience hypnosis if you can breathe and rest. In a state of naturally induced relaxation, the mind and body can work together to heal, mend, overcome, learn, create, and experience something wonderful.

If you are ready to experience a wonderful new learning experience, relax and get comfortable. In this session, I am going to teach you methods of taking physical control over your body. These methods can be used at any time and they have tremendous applications for managing many aspects of your life. You do not have to try to experience anything. You can let go and just enjoy this session in any way that you want to. Pain may be the awareness that caused you to seek out this hypnosis session, but after this brief introduction, the word "pain" will not be used again. You already know how to experience pain and you have been easily aware of your pain. In fact, you may have even been told by doctors or others that you would feel pain. As a result, you have been consciously expecting and even anticipating pain. You have no difficulty being aware of pain and so there is no need for me to focus on pain or even use the word during our session time together.

The skills in this session will not focus on pain, but rather on teaching mindfulness. You will learn that pain can coexist with comfort, that hurt can be reduced, and that it is the awareness of comfort that is important. So often we have been asked, "What is your pain level?" This question only serves to increase our awareness of pain. Rarely does anyone ask, "What is your comfort level today?" That is the question that I am going to focus on, because increasing comfort is the easiest way for the mind and body to work together to end pain.

You can focus on anything you want to and this session will teach you how to develop your skills of creating awareness of things you may have never been aware of before. Awareness is an amazing skill that is a function of the intuitive and creative part of the mind. As you relax, you may become aware of many things and this is good.

Induction

Awareness induction or eye fixation is recommended.

Deepener

Use your favorite deepener.

Suggestive Script

You have learned a lot about awareness up to this point, and it is in this state of relaxation that you may choose your level of awareness. You can focus on each word, relaxing and yet aware of your awareness, or you can drift into a deeper state of relaxation. Perhaps your conscious mind won't even pay attention to the specific words I use, but rather, simply absorb the experience through the lens of the subconscious mind. Either way is fine. This is your learning experience and you can experience it in any way that is meaningful to you. Perfect.

Just as you were able to easily use the creative part of the mind to shift your awareness, you can see how this skill has transformational power. This is because no matter how difficult experiences may seem, a person who is still alive has more right with them than wrong with them, even in periods of illness or discomfort. Up until this point, any treatments you have been receiving have served to draw your awareness to what is wrong. However, it is now time to shift your awareness to what is right within you. Even though it may have been a long time since you focused on what is well inside of you, you can now begin feeling a sense of that wellness.

Just like that moment when you moved your awareness inside of the mind and became more aware of the part of the mind that creates awareness, you can become aware of wellness inside of you, no matter how small that wellness feels. For some, it may feel like energy within. For others, it may be a visualization of wellness inside of you or just a comprehension of the concepts in this session. Either way is fine. For each person, this experience will be a little bit different. As you focus on the wellness inside, you can continue to learn and experience awareness. Focus on experiencing that within you that is well, the cells that are healthy, and the ease with which your blood can move throughout your body. It brings nutrients and the essence of life even to those parts of the body that others have labeled as sick.

It is remarkable how this new awareness is healing, in itself. It is something you can sense even if you have never felt this before, or even if it has been a long time since you were attentive to the wellness within you. As you relax deeper, allow yourself to let go of the negative projections or predictions of others. Just focus on this awareness of what is well within you and go deeper into yourself, creating visual images with the mind that bring a sense of serenity and peace. It is a wonderful experience, isn't it? This is not an experience that I have created, but rather it is an experience that you have created within yourself. This skill is one that you will easily be able re-experience at any point in life when you need to overcome the words of others or the predictions of those limited by their own previous experiences.

Again, allow yourself to hear each word I use with the subconscious mind. You can also choose to drift, dream, and float through this experience. As you experience this session, use the creative part of the mind to imagine two knobs. Imagine one on the left and one on the right. These knobs exist on a computer controller that is connected to your body and mind. In reality, such a controller exists in your mind, so this visual representation that I have asked you to create is quite practical. The knob on the left controls the volume of discomfort. Up to this point, you have been listening to it on high volume. As you visualize this knob, slowly turn it to the left to turn down the volume of discomfort. Feel the hurt fade away and experience your intuitive ability to turn down this level of awareness. Excellent.

Now imagine that the knob on the right controls your comfort level, which is something that you have most likely not been listening to on a daily basis. Turn this knob to the right and experience the increase of comfort within you, as the feelings of peace and serenity increase. Feel the sensation of wellness and physical comfort increase. This is not just a mental exercise. You can actually experience these wonderful changes and perhaps you have already noticed an increase in the heaviness of your body as you enter a more peaceful state of relaxation. Or, perhaps you notice lightness in the body as the volume of comfort energizes your cells.

Some people notice other sensations such as a tingling in the top layer of the skin. Still, other people do not have such obvious changes as they learn this process. Rather, they experience change over a period of time as they practice the skills of hypnosis. Again, any outcome is okay because this process of creating new awareness is an experiential process. It is one that you will experience as the wellness within you becomes the awareness that you allow to spread through every fiber of your being.

Everyone who engages in this process of increasing comfort also finds that as the knob representing comfort is increased, depression, despair, and anxiety give way to new feelings of release, acceptance, healing, and hope. A new freedom and a new peace is gained by being attentive to comfort. Increasing that comfort and wellness that is already inside, will easily bring these positive feelings to the attentive mind and carry you easily through each day.

Sometimes when I am teaching people these skills, they fear that the feeling of wellness or comfort that they have been able to increase is temporary or only exists for the length of the hypnosis session. However, this is not true. You can easily create the experience of turning down the knob on the left and decreasing the volume of hurt or discomfort at any time. In fact, you can even do it again now by visualizing that control on the left and decreasing the intensity or volume of discomfort down to barely a whisper. It feels wonderful, doesn't it?

Now, in the stillness of the mind, slowly turn that imaginary knob on the right to a higher level of comfort. Notice that as you slowly turn it to the right, you become more aware of your comfort level. Feel a sense of wellness swelling within you and the energy of health and a spirit of hope. Excellent, you are doing great. You can turn it as high as you would like, easily increasing your comfort to the highest level that you have experienced in a long time. What an amazing feeling of freedom this brings. Enjoy this experience for a moment and simply allow the mind to expand the awareness of your ability to be attentive to the health and wellness within you. It feels remarkable, doesn't it?

You have the ability to recreate the feeling you have now, by just closing your eyes for a brief moment and bringing yourself back to this point. In fact, you will find that in the mornings, you will easily begin to turn the control knob and increase the volume of your comfort. You can carry this positive feeling with you throughout all of the tasks that each day brings. Wonderful.

Although your awareness may shift to difficulty from time to time, you will easily recognize that inside of you is a place of wellness than can be focused on as easily as any other state. You will listen to the body and respond by caring for it through proper nutrition, rest, and even various treatments such as medication or medical care. However, as you do, the awareness of wellness will remain paramount, increasing each day as you master these skills and move forward in your life.

As you relax, having enjoyed this period of rest and rejuvenation, you can continue to keep your eyes closed and be aware of the experience within you that brings wellness to every cell of the body. Be aware of how the heart works with your lungs and then with your blood supply. Be aware of how the blood supply works with the muscles of your body by increasing comfort in every muscle, tissue, and cell. It is a wonderful feeling.

I am going to give you another moment to experience the stillness of this spot and whatever else you need to experience at this time. Know that it is perfectly okay to shed a tear of releasing past sorrows or even experiencing new joy. At this time, let go of any discomfort and increase the experience of comfort, peace, and serenity. Or you can just be still and enjoy the sounds of life. Wonderful! Enjoy the peace, serenity, wellness, comfort, wholeness, strength, and power.

From this point forward, anytime you see the color blue, it will remind you of the skills that you have learned in this session. In fact, it will be a calm and cool reminder of the healing forces that exist within your body and mind. Although there is no real reason, the color blue will simply become a comfort to you by strengthening your experiences and increasing your comfort level. You might notice a blue car driving by, a blue cup, a blue shirt, or even the blue sky. You will notice blue in each and every place in a way that you haven't before. This increased awareness of blue will be a demonstration to you of your ability to sharpen your awareness of anything, including increases in your comfort level. In fact, blue will bring a smile to the face as hope, comfort, and restoration are clearly within your reach.

As you continue to keep your eyes closed, it is time to reorient to the room around you. Shift your attention to the part of the mind that creates awareness. Begin feeling more awake and alert. Now, shift your awareness back to that point directly in front of your closed eyes, knowing that this time has been beneficial and finding it even easier in the future to experience hypnosis, healing, and health.

Now, imagine that spot in the center of the room. Open your eyes if you would like to, and feel more energetic. Stretch out any muscles, take in a deep breath, and focus your attention at the starting point on the far wall. Good! Breathe in deeply, restoring all energy to the body and stretching the back and neck. Now you become fully alert, fully awake, and ready to practice these healing principles in all of your life experiences.

Overcome Insomnia and Sleep Better

Evidence

Effects of relaxation training on sleep quality and fatigue in patients with breast cancer undergoing adjuvant chemotherapy. (Meral Demiralp, Fahriye Oflaz, Seref Komurcu) Journal of Clinical Nursing Volume 19, Issue 7-8, pages 1073–1083, April 2010.

Aim: The purpose of this study was to investigate the effect of progressive muscle relaxation training on sleep quality and fatigue in Turkish women with breast cancer undergoing adjuvant chemotherapy.

Background: Sleep problems and fatigue are highly prevalent in patients with breast cancer. Progressive muscle relaxation training is a promising approach in ameliorating the sleep quality and reducing the fatigue associated with cancer and its treatment.

Design: A prospective, repeated measures, quasi-experimental design with control group.

Methods: The study sampling consisted of 27 individuals (14 individuals formed the progressive muscle relaxation group, 13 individuals formed the control group) who met the criteria for inclusion in the study. Progressive muscle relaxation training was given to the progressive muscle relaxation group, but not to the control group. The effect of the progressive muscle relaxation training was measured at different stages of the treatment. A data collection form, the Pittsburgh Sleep Quality Index and Piper Fatigue Scale, was used to collect the data for this study.

Results: The progressive muscle relaxation group experienced a greater increase in improved sleep quality and a greater decrease in fatigue than the control group.

Conclusions: The findings indicated that progressive muscle relaxation training would improve sleep quality and fatigue in patients with breast cancer undergoing adjuvant chemotherapy.

Relevance to clinical practice: Progressive muscle relaxation training given by a nurse can improve sleep quality and fatigue in patients with breast cancer. It is important to start relaxation training just before chemotherapy to decrease the frequency and severity of sleep problems and symptoms such as fatigue during chemotherapy.

Note to the hypnotist

Sleep difficulties can be due to a number of factors. This script deals primarily with anxiety as a source of those difficulties. However, clients should be given a full assessment as to the nature of sleep difficulties, including a medical evaluation.

Pre-Talk

I can't think of a drug, face strip, or noise control device that can help a person sleep better than hypnosis. This is because hypnosis actually teaches a person how to sleep. "Hypnos" is the Greek word for sleep and although a person in hypnosis isn't actually asleep, the process for entering trance is the same process as drifting off into a deep and healthy sleep.

This session is designed to teach you how to sleep. Perhaps your sleep difficulties come from sensitivity to light, noises, stress, or even your sleep partner's movements. It is amazing, but the mind actually has the ability to learn to tune out these distractions and even use them as sleep cues to reinforce healthy sleep patterns. Perhaps anxiety or racing thoughts have kept you from relaxing and drifting into a deep sleep. Hypnosis is a teaching process that can help you manage anxiety, turn down the volume of racing thoughts, and sleep well.

Obviously, issues like jetlag, uncomfortable beds, nicotine or caffeine addiction, and physical illness are temporary or changeable discomforts that we have control over. Hypnosis can even help in these situations by training your eyes, your lungs, and your mind to let go and enter into a deep, restful sleep.

Some try to use drugs or alcohol as sleep aids. The problem with this is that they can disrupt the natural REM cycles that are necessary for healthy sleep. Hypnosis is empowering and will help you to sleep all night and wake up feeling rested.

Induction

Progressive muscle relaxation is recommended.

Suggestive Therapy

Although you are not asleep, you have taken the first step in relaxing, and relaxation in itself can help a person to feel rested. You can use this skill to manage anxiety or relax the body and begin the process of drifting into a deep and profound level of sleep, in a few minutes. You can also use this state of hypnosis to shift your awareness from external distractions to a point of perfect peace and serenity. The mind is quite powerful.

You may be aware of many experiences at this point, including the room temperature, sounds from outside, and others nearby. In the past, these may have been cues to distract you, but now you can use this new knowledge to recognize your ability to shift your awareness to any experience, including those you create in your mind. The mind has a place deep inside of it where awareness is created. Until hearing this, perhaps you were unaware of this place. However, now that you know about it, you can shift your awareness from all that is outside of you, to this place in the mind.

From this new vantage point, you can use the creative part of the mind to begin to drift, dream, and float inside of your own awareness. You can use the creative part of the mind to envision yourself, perhaps under a clear blue sky. Perhaps you are in a beautiful place you have been to before, a place that you would like to go, or even a place you have created that is relaxing and peaceful. Imagine that as you look to the clear blue sky, you can see a single, white, puffy cloud, gently floating by. See the soft edges of the cloud and the pure energy of its radiance. Watch as it gently moves from one side of the sky to the other. Imagine that all of your stress, discomfort, or concerns become enveloped by the energy in this cloud.

As the cloud moves into the horizon, becoming smaller and smaller, you become increasingly aware of your mind's ability to relax and create its own experiences. It has long forgotten about any difficulties that used to concern you. As it drifts into the distance, it will eventually become so small that it completely disappears. At this point, you are sensing an awareness of your feelings of deepening relaxation. You feel yourself moving twice as deep into relaxation of mind and body. Go all the way down, into deep mental relaxation. Perfect. With a sense of security and comfort, you can be happy about the new experiences and lessons you have had that can help you to sleep deeply at night.

I am going to give you some suggestions now. You can pay attention to each and every word with the conscious mind, or you can drift, dream, and float. You can hear all of the words with the subconscious mind and only every other word, every fifth word, or every tenth word with the conscious mind. Perfect. These are suggestions that you have asked me to make when you asked me to help you sleep better, so they are actually suggestions that come from you. You will find it easy to internalize and experience these suggestions now and at any time you need a restful night of deep tranquility.

Each night as you experience self-hypnosis, you will find it even easier to follow the instructions and drift more deeply into trance and eventually into sleep. As you listen to your breathing and perhaps think about the sweet, soft sounds of nature, you will begin to move your thoughts from those that are external to those that are inside the part of the mind where creativity and awareness is created. This will allow you to detach from anything that previously hindered your sleep.

As you continue to relax, let yourself freely and lazily double the sensation of relaxation, in both mind and body, letting go of any tension in any muscle. This feeling of the loosening in the muscles will bring you even further into trance. You know that at this moment, you could open your eyes or wake up entirely, but it feels so good to finally let go that you will continue to do exactly what you are now doing and experience an even deeper level of hypnosis or relaxation.

Although you know you have the ability to awaken if something needs to be attended to, at this point there are no outside concerns. Therefore, it is okay to let yourself go deeper and deeper into trance. In the morning when it is time to take yourself, your internal awareness of the freshness of a new day will bring an energy and sense of wellness that will carry you through tomorrow. It is wonderful to feel refreshed and revived from a good night of sleep. However, this time is not now and it is perfectly okay to continue focusing your awareness inward and eventually drifting from hypnosis and into deep sleep.

As I count backwards from ten to one, enjoy the feeling of heaviness in the arms and legs, any feelings of coolness or warmth you desire, and the sensation of rest in both mind and body. Ten, rest . . . Nine, perfect . . . Eight, peace . . . Seven, slowly drifting into sleep . . . Nine . . . Eight . . . Seven . . . Six, with each number not knowing where sleep begins . . . Seven . . . Six, just experiencing the process . . . Five, good . . . Four . . . Three, total comfort . . . Four, safe . . . Three . . . Two, wonderful . . . One, drifting to a point where awareness of awareness simply drifts off into the distance . . . Sleep . . .

Loving-Kindness Meditation

Get into a comfortable, quiet place and simply breathe in and out. You don't have to breathe in any special way. Close the eyes. With the eyes closed, as you breathe in and breathe out, think of a person in your life who has been a teacher or a mentor. It could even be a person on a national or global level who you really admire. Think of that person who has given so much to our world or to you and as you breathe in and out in this moment, allow yourself to feel compassion and love and respect for that person. Maybe it is an historical person who is no longer with us or someone who has been an important mentor in your life. Either way is fine, of course. But right now, as you breathe in and out in this time and this place, allow yourself to experience a feeling of love and kindness towards that person. It's pretty easy to do since they've given so much.

As you breathe in and out, in this next step, think about a person who is important to you who is from your world. Perhaps it is a parent, a close friend, or a boss. Perhaps it is somebody who has touched you on a personal level in some way. Maybe it is a child, a neighbor, or somebody who you fellowship with on a regular basis. As you breathe in and breathe out, breathe in love and exhale kindness. In your own mind, direct that feeling towards them. You might even notice a smile on your face as you breathe in love, exhale kindness, and direct your energy in your mind towards this person.

Now, as you breathe in and breathe out, think about a person in your world who you might not have ever taken the time to think about before. Maybe it is the mailman you haven't spoken to, who merely drives by your block, or the bag teller at the bank that you see once or twice a week, or the person who takes your clothes when you drop off your dry cleaning. Maybe it is someone who you exchange pleasantries with, but have never really talked to or paid too much attention to before. Think about this person, and as you breathe in and breathe out, allow yourself to channel the energy of love and kindness towards them. Just take a moment to appreciate them, simply because they are who they are and they are a part of your world. Notice then, that by taking the time to do this, you've already expanded your capacity to create and experience love and kindness and share that with those around you.

To this point, I've given you three tasks. Each one was different, but each one was easy to do. However, now as you breathe in and breathe out in this moment of mediation, think of a person in your life who perhaps has been difficult to have as a part of your world. Think of a person who you might have fought with, conflicted with, or maybe even avoided. Maybe it is just someone you find to be extremely difficult to be around. Perhaps think of a person who you do not even like or someone who you feel very strongly about. It may even be a person that you hate.

In this moment, as you breathe in and breathe out, you can recall how you were just able to direct love and kindness towards that leader in your community, that mentor in your personal life, that friend, or even that acquaintance who you hadn't thought too much about before now. Just take that energy of love and kindness and direct it towards this person whom you've had a difficult time accepting as a part of your world. Notice that because we are in the practice of cultivating love and kindness, it's as easy in this moment to direct love and kindness towards this fourth person, as it was to all the others.

With the next breath and before you open your eyes, breathe in and let that breath energize you, reaching every cell in your body. Now, open the eyes and feel fantastic, wide awake, and refreshed from this love and kindness meditation exercise.

Developing a Wealth Mindset

Pre-talk

It's a lot easier to develop a wealth mindset when we begin to notice wealth rather than noticing lack or poverty. I've never met anyone who has no strengths and no resources. Furthermore, our strengths and our resources are often what we use to generate wealth, from this moment forward. The easiest way to begin noticing wealth is to practice noticing anything. This is really one of the foundations in meditation, self-hypnosis, or really any self-actualizing technique.

Meditative Experience

So right now, where you are, just close your eyes for a minute and breathe in and breathe out. As you continue to breathe in and breathe out, notice the breath. Now, the breath is a great thing to focus on because no matter where we go, we always have our breath with us. The art of cultivating awareness of what we have within can really begin by simply noticing the breath. As I've told many clients, as long as you are breathing, you're actually okay.

So, you have your breath right now. It's something that you do have already. Pay attention to it, study it, and become aware of what it's like to breathe in, what it's like for the air to travel deep into the body, and what it's like for that breath to turn into an exhale. Just practice paying attention to your breath, observing and noticing the rate at which you breathe and how the way you breathe feels. You can focus on the various sensations of breathing, such as which nostril is breathing in the most air and what it feels like to breathe out. Just take a moment and continue to notice the breath.

When we pay attention to anything, we practice the skill of paying attention to everything, and this is really important. Right now, as you pay attention to your breath, pay attention to your wealth. Maybe it's not yet measured in dollars, gold, or stocks. Although it might not be measured in those accruements of wealth, notice the wealth that is within you. Is it potential? Is it creativity? Is it ideas? What wealth do you already possess within you? What strengths do you have? Are you trustworthy, loyal, courteous, kind, cheerful, thrifty, clean, brave, or reverent? What strengths do you have? What's awesome about being you? Feels pretty good to notice those things, doesn't it?

What resources do you have? Maybe you don't have a fancy high-dollar European import to drive around in, but do you have access to a vehicle or public transportation? Perhaps you're not eating at luxurious restaurants each and every night, but do you have your most basic needs met? By measuring just those things against the world standard, I can almost guarantee that you are in at least the top half of the world's wealth.

So, how do you move from that top half to that 1%? One key is being mindful of what we have and by practicing this type of mindfulness meditation, your awareness of all of the strengths, resources, and ideas you have will increase. As you focus your attention on those things and your awareness of them increases, they grow, expand, and begin to demonstrate themselves in your life. That is what we call a universal truth . . . that which we focus on, expands. Now that you are focusing on what you have, those things will continue to increase in your life. Very good. Now, just open your eyes and take a deep breath, having enjoyed this experience and feeling wonderful.

Increasing Energy Levels: Breathing in Energy

Evidence

Fatigue during breast cancer radiotherapy: An initial randomized study of cognitive– behavioral therapy plus hypnosis. Montgomery, Guy H.; Kangas, Maria; David, Daniel; Hallquist, Michael N.; Green, Sheryl; Bovbjerg, Dana H.; Schnur, Julie B. Health Psychology, Vol 28(3), May 2009, 317-322.

Objective: The study purpose was to test the effectiveness of a psychological intervention combining cognitive–behavioral therapy and hypnosis (CBTH) to treat radiotherapy-related fatigue.

Design: Women (n = 42) scheduled for breast cancer radiotherapy were randomly assigned to receive standard medical care (SMC) (n = 20) or a CBTH intervention (n = 22) in addition to SMC. Participants assigned to receive CBTH met individually with a clinical psychologist. CBTH participants received training in hypnosis and CBT. Participants assigned to the SMC control condition did not meet with a study psychologist.

Main Outcome Measures: Fatigue was measured on a weekly basis by using the fatigue subscale of the Functional Assessment of Chronic Illness Therapy (FACIT) and daily using visual analogue scales.

Results: Multilevel modeling indicated that for weekly FACIT fatigue data, there was a significant effect of the CBTH intervention on the rate of change in fatigue ($p < .05$), such that on average, CBTH participants' fatigue did not increase over the course of treatment, whereas control group participants' fatigue increased linearly. Daily data corroborated the analyses of weekly data.

Conclusion: The results suggest that CBTH is an effective means for controlling and potentially preventing fatigue in breast cancer radiotherapy patients.

Note to the hypnotist

For this practice, have yourself or your client sit in an upright position, if possible. Sitting up in a chair or on the side of the bed are both good options. Make sure you or your client is stable, with feet squarely on the floor and back and spine in alignment. Often, just sitting in such a position and allowing increased oxygen because of aligned posture, brings a sense of increased energy. Of course, if it is not possible to sit in a chair, a reclined position where the spine is in alignment will do because this posture helps one breathe in more fully.

Pre-talk

Many people ask me how hypnosis can help with fatigue. After all, isn't hypnosis like sleep? Although hypnosis uses relaxation and can be used to help a person sleep, relaxation can also be energizing, especially when we take mindful breaths which oxygenate the blood supply and help us to become healthy and energized.

Suggestive Therapy

I know that you have been breathing your whole life, but take a moment and pay attention to your breath. Notice where the air enters the nostrils and what it feels like to breathe in and breathe out. You might even notice that in this position of alignment, breathing is a bit more satisfying. With the next breath, intentionally breathe in deeply. Take in as deep a breath as possible, breathing in and out, and observing the breath as you do. Did you notice something? Did you notice that as you breathed in, you tightened up the shoulders and neck and even puffed up the chest a bit? I don't know why we do this, but often when we take a big deep breath, we puff up the chest and tense up those muscles. This actually limits our breath, letting air into only the top part of the lungs.

I am going to teach you a new way of breathing to energize the body by bringing oxygen to the lowest part of the lungs and filling the blood supply with a blast of oxygen that can help you to energize, anytime you feel fatigue. Close your eyes and as you breathe in and breathe out, pay attention to your belly. In fact, you can even rest your hand on your belly during this exercise if you want to.

With your body in alignment and your eyes closed, imagine that inside of your stomach is an empty balloon. Imagine that balloon is inside of your belly and empty. Now, a new way to take in a deep breath is to not to puff up the chest, but to fill your lungs by imagining that this balloon in your belly is being filled with air, breathing from the diaphragm, rather than the chest. Take in a deep breath now, filling that balloon with air and exhale by just letting the air out of the balloon.

Open the eyes. Do you notice a difference? Try it again, with the eyes open or closed. Take in another big breath, breathing with the diaphragm as if you are filling that balloon with air. Now exhale. Do you notice your neck and shoulders do not tense up? Do you notice how air goes to the deepest part of the lungs? You will also notice that as you take in a deep breath or two, you can feel a sense of energy throughout the body, countering that feeling of fatigue you may have felt earlier. You can do this, at anytime and anyplace when you need to rejuvenate from the fatigue of treatments, medications, or even just to promote health and wellness.

Now continue to breathe in and out, trying neither to speed up nor slow down the breath. Just breathe with a calm, smooth pace. You have dedicated this time to learning something new and you have. In this state of concentration and learning that you have created, you can close the eyes again and enter into that calm state of hypnosis, where you can practice affirmations. Affirmations are a way of energizing the mind by countering self-talk that may be contributing to fatigue.

Often when we notice symptoms of fatigue, or even in

anticipation of certain medications and treatments, we might tell ourselves things like, "I know I will be tired," or "That always makes me feel worse." While in the moment these thoughts may appear to be true, we really do not know what the future holds. We know that at different times we can feel different feelings and taking a moment here to focus on the present and the newfound ability you have to take deep breaths that energize you, you can also tell yourself affirmations. Tell yourself, "My breath can energize me. My breath can energize me. My breath can energize me." You can even tell yourself, "With every breath, I bring energy to my body. With every breath, I bring energy to my body. With every breath, I bring energy to my body."

You know that even if it is hard at times, it is important to choose a small healthy meal, even when food is unappealing; because the nutrients can help you gain strength and energy. When you feel a sense of fatigue, ask yourself, "Have I eaten today in a way that will help me gain energy?" Sometimes a quiet time like this is a place to remind yourself of something else you already know, and that is that physical activity brings strength and healing. So, even if it is difficult, you have probably been given instructions for stretching, walking, or movement from your physician or physical therapist. It is important to follow these recommendations, because once you begin to create energy through movement, it increases the energy by compounding it and combating any fatigue you feel.

You have done great today in this short session, learning a new strategy for breathing and reminding yourself of ways you can counter fatigue. You have the power to do this now and I know this is true because you have taken the time to participate in this session and will take action on these suggestions, improving in every day and in every way.

Awakening

Use your favorite awakening here.

Future-Pacing Financial Success

Pre-talk

There is an interesting quote about life and death from Frederick Nietzsche. The quote is very interesting and I'm paraphrasing, but it basically said, "*If you were to die tomorrow and you discovered that the afterlife was you living your exact same life over again – each moment, each joy, each difficulty, each triumph, each tragedy, each experience that you have lived in this life, but had to live it over and over and over again, would that be heaven or would that be hell?*"

It's really a powerful question that causes a person to think. If you're thinking to yourself that living the exact same life that you're living now over again would be misery, then it's time to future-pace and change that script. Even if you do think that it would be heaven to live your life over again, I think we can still future-pace and make improvements and make things better and to experience an even greater level of success by expanding our expectations. So, let me guide you through a short process of future-pacing. I think it's important for us to realize that in the now, we can create a future, rewrite our life scripts, and live abundantly.

Suggestive Therapy

Take a minute and just breathe in and breathe out. As you breathe in and breathe out, pick a point over on the far wall and bring all of your attention to that point. As you breathe in and breathe out, take this moment and this time to learn something new and to experience success. Bring all of your attention to that point over on the far wall and let that point be the center of a movie screen. Imagine that a movie of your life is playing and see and image of yourself, as you are right at this moment. Now, let that movie screen change and let it show a picture of you, tomorrow. See yourself happy from learning new things in this session. See yourself tomorrow, implementing new ideas. See yourself maybe even planting a wealth garden with actual flower pots and putting in that first dollar.

It's easy for you to create that image and see yourself in that movie screen on the far wall, isn't it? Now, extend that out a week. Imagine that you've been planting dollars for a week. You've been setting goals for a week. At the end of the week, you've experienced some success. See yourself and notice how you might feel in that movie of you, at that point in time, on the far wall.

Know that when you apply the principals of success in every aspect of your life, life is magnificent. So, see yourself as you know you'll be a year from now. Maybe you are stepping into that vision board you've created. You can even see yourself, as if this is a movie of your experience, with the people that you love, the places that are important to you, and the things that you use to measure wealth ten years from now, twenty years from now, and maybe even thirty years from now. In fact, you can even step into your old age and look back and review where you started and where you came from.

Feels pretty good to future-pace, doesn't it? There's probably even a smile on your face as you view that screen on the far wall. As you breathe in and breathe out, enjoy that image. Enjoy that picture you've made of what the future is like, as you create success and empowerment in every area of life that is important to you.

Now, with the next big breath, just breathe in and breathe out. Let that oxygen go to every cell of the body and let it energize you. If for some reason your eyes are closed, open the eyes. If your eyes are fixated on that point, move your eyes. Let a smile come across your face and know that by creating with your mind an idea of the future, you've allowed yourself to rewrite the scripts that once held you back, turning them into the scripts that will now allow you to start a new chapter of life that's abundant, full, and rich.

Stress, Mood, and Anxiety

Evidence

Hypnosis in the treatment of anxiety- and stress-related disorders. D Corydon Hammond. Expert Review of Neurotherapeutics February 2010, Vol. 10, No. 2, Pages 263-273 Self-hypnosis training represents a rapid, cost-effective, non-addictive, and safe alternative to medication for the treatment of anxiety-related conditions. Here, we provide a review of the experimental literature on the use of self-hypnosis in the treatment of anxiety and stress-related disorders, including anxiety associated with cancer, surgery, burns and medical/dental procedures. An overview of research is also provided with regard to self-hypnotic treatment of anxiety-related disorders, such as tension headaches, migraines, and irritable bowel syndrome. The tremendous volume of research provides compelling evidence that hypnosis is an efficacious treatment for state anxiety (e.g., prior to tests, surgery and medical procedures) and anxiety-related disorders, such as headaches and irritable bowel syndrome. Although six studies demonstrate changes in trait anxiety, this review recommends that further randomized controlled outcome studies are needed on the hypnotic treatment of generalized anxiety disorder and in documenting changes in trait anxiety. Recommendations are made for selecting clinical referral sources.

Pre-talk

This session is designed using the principles of self-hypnosis to assist you in ending the feeling of anxiety, anger, and depression. Many people who use the principles of self-hypnosis or meditation find that by taking a moment in the middle of life's turmoil to re-center themselves is a great way to relax. You might even wonder if something so simple can really help you. The answer is YES! There is hope for ending depression and it comes from the desire to be happy that is already inside of you. You can find calm from anxiety, and you can even give up stress, anger, or frustration by practicing these simple ideas.

Induction and Suggestive Therapy

Use your imagination and envision a peaceful waterfall, focusing on your ability to create such a visualization. If you cannot picture a waterfall, imagine what the gentle water feels like, or even what it sounds like. Focus your attention on the moving water. I love water. For some reason, watching water move brings me a sense of peace. Perhaps it is because moving water represents the power of life, or perhaps the movement simply distracts me from stress, but either way I find every time I watch water move, feel it, or listen to it, I feel calm.

As you relax, scan your whole body. Anywhere you are holding on to the tension of the day, let those muscles become loose and relaxed. Often we don't even realize where we are carrying tension until we notice it and make the conscious choice to relax those muscles. Focus all of your attention on an image that you create of moving water. As you imagine the water flowing, let any immediate stressors or anxieties flow with the water from you and to a far and distant place.

Imagine for a moment, as you focus on the moving water, the feeling of not only the water moving, but also your stress moving with it. Many will find this easy to do, but others will find this not an easy thing to do. Perhaps this is because it has been a part of life for so long, that letting it flow is something that takes time. That is perfectly okay.

Notice how your body has become more relaxed. After focusing your attention for so long, your eyes might even be tired or heavy. It is perfectly okay to let them close and use this time to re-energize not only the body, but the mind and soul also. Go ahead and close your eyes, using your imagination to drift into the scenery of our waterfall.

Pay attention to your breathing. Notice how it has become slow, smooth, and rhythmic? This isn't even something you have tried to do, but something that comes naturally by taking a moment for yourself and letting go of any obvious tension. As your hands rest on your lap, notice the sensation of relaxation and calm you have already achieved. Now say to yourself the word "warm" and focus on your hands, letting them feel a sense of warmth. Say to yourself, "My hands are warm," and as you do, notice the sensation you have created of warmth in those hands.

Now say the word "heavy" and notice how heavy your resting hands are. Say to yourself, "My hands are warm and heavy. My hands are warm and heavy." Now focus on your feet. Say to yourself, "My feet are warm and heavy. My feet are warm and heavy." It is amazing that as you say "My feet are warm and heavy," you can begin to feel that sense of warmth from within and that sense of heaviness. This is the first lesson and that is that we can control the way that we feel. This is true both physically and mentally, no matter how difficult life situations are. That sense of warmth and heaviness also brings a sense of relief from the weariness of life, letting you recharge the mind and body.

Focus on your breathing, noticing it smooth and rhythmic. Relax even deeper with each breath. Although it may feel magical to relax this deeply, especially since you haven't felt this calm in a long time, this is a totally natural state. It is a state that you have created, rather than me. It is also a state that you can re-experience at any time and I will teach you how to do this in a moment. However, now is a time to focus on your desires, being free from panic or anxiety, by feeling a state of calm. Notice how the sense of heaviness is like a calming anchor, allowing you to feel physically calm, even if your mind might wander or race. Perhaps you desire hope rather than depression, and through this simple exercise of creating warmth and heaviness, you can see how you not only have the ability to create physical sensations, but also feelings of hopefulness.

Water always makes me feel hopeful. I do not know why, but it does. Perhaps it is because it reminds me that nothing stays the same or perhaps because its power and energy revitalize my spirit. Do you desire freedom? Do you desire success? Take a moment and create an affirmation that reflects your heartfelt desire. You can say something like, "I am free from stress," or "I can create hope from inside," or you can even focus on a single word like "happiness", "calm", or "forgiveness". As you listen to the water in the background, take a moment to focus on this word or on your affirmation. Repeat in your mind this word or affirmation, seeing the letters as you do, spelling it out, and hearing yourself speak these words and noticing the feeling that these positive words or affirmations bring. It really is amazing that how simply saying something manifests it as a reality. However, remember everything that is or ever was, had to be a thought first.

To be hopeful, to be calm, or to be forgiving, you must begin with a thought and then you can truly experience this more and more each and every day. This thought is like a rain that fills a stream and leads the river, resulting in a powerful waterfall. A great way to reinforce it is to write this affirmation on a card or sticky note and tape it to the bathroom mirror, your dashboard, or to the monitor on your computer. Let it be a constant reminder, despite other factors in life, of your ability to turn thought, which is mental energy, into emotions and success.

Now relax even deeper, noticing how remarkably easy it has been to set aside a few moments to re-energize. Even though life requires action, you now have a starting point for re-energizing during difficult times and a starting point for creating thoughts which turn into results. In fact, you can even congratulate yourself for taking the time to invest in your success, knowing that by starting this process, hopefulness, calmness, and happiness are the result.

As you continue to relax, pay attention to your right thumb and index finger, touching them together as if you are making an "okay" sign. As you feel a sense of calm, touch them together, pressing firmly. This is associating this feeling with this action. Anytime over the next hour, day, week, or even year, whenever you are stressed, anxious, or down, touch those fingers together, allowing yourself instantly to recall and re-experience this state you have created. This state is one of calmness, hopefulness, and success. You can do this at work, at home, or in any difficult situation. Instantly, even though it seems amazing, you will bring yourself back to this starting point of serenity and creativity, allowing any temporary stress to pass without escalating.

As we near the end of this session, you can open your eyes at any point, or you can keep them closed for another moment or two while you feel the floor below your feet and the air in the room around you. Let yourself feel a sense of energy in the muscles of your body and in your spirit, feeling refreshed, energetic, and ready to be positive in any situation.

Pre-surgical Preparation

Pre-talk

The evidence shows that those who learn methods of hypnosis, including visualization and relaxation techniques prior to surgery, can reduce complications, feelings of illness, and promote physical healing. I think it is important for people to know that there is hope for decreasing distress during times of surgery, and hypnosis is one way to increase your feelings of wellness both before and after surgery.

How does hypnosis help? We know that hypnosis teaches skills which promote healing, and by learning strategies for minimizing discomfort, you can even make it through a complex surgery with a greater level of comfort.

This is not a long session. It is designed to give you a few moments to be mindful of the present and to detach from any anxiety or fear. It is also a tool for healing, using visual imagery to promote health.

Induction

Begin by taking in a breath and closing the eyes. Just breathe. There is no special way to take this breath. Just let the eyes relax as you breathe in and out. Gently bring your attention to the solar plexus, which is the bundle of nerves in the core of your body that is behind your stomach and lower than the sternum. In personal fitness training, this is often called the center of our gravity. In metaphysics, it is right above the third charka on the spine. Although it is a place of mystical attributes, the role of these nerves and sensations radiate and extend to every part of the body in scientific literature as well. Do not worry about its exact location at this point, because you are close enough to begin noticing how the breath moves the diaphragm, how it contracts and expands, and you can feel the movement of the organs within you in this area. Perhaps you are breathing "into" or from the chakra in the spine, or noting the sensations in and around the stomach.

Each person will be aware of different sensations in these areas, and each time you practice this meditation, you will probably note new sensations and experiences. Mindfully observe these sensations, and with a curiosity growing stronger with each breath, bring all of your awareness to this area. Notice things such as the weight of your clothing, the depth of your breath, or the energy through these sacred and interconnected places of the abdomen and solar plexus.

It is also important to note that you have just practiced something important by moving a mindful awareness of one thing, to a mindful awareness of something else. Mindfulness is not about only one awareness, but rather a discipline of being able to shift mindfully from moment to moment and between various states in life.

Continue to breathe in and out, and mindfully observe the solar plexus and the nearby muscles, organs, and bones. If at any time you notice a sensation or awareness that is uncomfortable or even distressing, do not judge that awareness. Rather, practice being good to yourself. Without following that judgmental or harsh thought, just bring your attention back to this place.

Right leg

When you are ready to explore more of your body in this moment, take in a breath. Inhale and exhale, following the breath as it goes all the way out, down through the core of your body, and follow the breath through your right leg. Imagine the breath flowing all the way into the right foot and each toe. By doing so, you have moved your awareness in this moment from the belly, all the way to your toes. You can even wiggle them, noting all of your toes being independent from the other toes. Note the sensations in the toes, the bottom of the foot, and the arch of the foot. As you continue to breath all the way into your foot, curiously observe the foot without judgment, including the bones and muscles all around it. Notice the weight of the shoe on your foot or the feeling of the air on your bare foot. Follow your awareness into the ankle and the bones of the ankle.

The practice here is simply to note feelings, sensations, and thoughts, while avoiding following those things. It is perfectly okay if a particular awareness is uncomfortable, difficult, or even sad. It is also okay if you notice strength, comfort, and relaxation. The point of the practice is to note these things. We can note them and accept them nonjudgmentally as they are, without becoming enmeshed in these thoughts and feelings. You can label them rather than follow them by saying to yourself, "That is a comfort," or "That is a pressure." Just breathe into the foot, observing in this moment, neither regretting the past nor fearing the future. Right now, this moment is just about your toe, your foot, your ankle, and just allowing this time to be as it is.

Continue to practice directing your breaths into your lower leg, the shins, calves, and even the knee of the right leg. Become aware of the texture, the long hair or the razor stubble, any scars or blemishes, and the feeling of the muscles. With each breath, direct more of your awareness to the lower leg. Again, you do not have to be hard on yourself if you find that by studying this part of the body, you begin to think of unrelated thoughts or have other feelings or sensations. When you notice them, just label them and let them be what they are without following them. Then, simply return to your mindful practice, with each breath, of simply observing and noting the lower leg and knee.

As you breathe in again, direct this breath to the thighs and the large muscles under the leg. Notice the weight of the clothing on your body and where these muscles are supported by the surface beneath you. Of course, our legs are held in place by the hips and the bones of the pelvis. Much of our daily work and exercise is in the large muscles of the buttocks. If you find your mind wandering and following an unrelated thought, like that it's silly to focus on the buttocks, or even that the word itself is silly, use that as a cue to mindfully return to your study of these regions and practice bringing your awareness back to the sensation of resting, the weight of your clothes, and each breath that directs you into this part of the body scan meditation.

Left leg

When you are ready to explore more of your body in this moment, take in a breath. Inhale and exhale, following the breath as it goes all the way out and down, through the core of your body. Follow the breath through your left upper leg and into the lower leg, imagining the breath flowing all the way into the left foot and each toe. By doing so, you have moved your awareness from the belly, all the way to your toes. You can now even wiggle them, noting all of your toes are independent from the other toes. Note the sensations in the toes, the bottom of the foot, and the arch of the foot. As you continue to breathe all the way into your foot, curiously observe, without judgment, the foot, including the bones and muscles all around it. Notice the weight of the shoe on your foot or the feeling of the air on your bare foot. Follow your awareness into the ankle and the bones of the ankle.

The practice here is simply to note feelings, sensations, and thoughts, and just avoid following them. It is perfectly okay if a particular awareness is uncomfortable, difficult, or even sad. It is also okay if you notice strength, comfort, and relaxation. The point of the practice is to simply note these things. We can note them and accept them nonjudgmentally, as they are, without becoming enmeshed in these thoughts and feelings. You can label them rather than follow them by saying to yourself, "That is a comfort," or "That is pressure." Just breathe into the foot, observing in this moment, neither regretting the past nor fearing the future. Right now, this moment is just about your toe, your foot, your ankle, and just allowing this time to be as it is.

Continue to practice directing your breaths into your lower leg, the shins, calves, and even the knee of the left leg. Become aware of the texture, the long hair or the razor stubble, any scars or blemishes, and the feeling of the muscles. With each breath, direct more of your awareness to the lower leg. Again, you do not have to be hard on yourself if you find that by studying this part of the body, you begin to think of unrelated thoughts or have other feelings or sensations. When you notice them, just label them and let them be what they are, without following them. Then, simply return to your mindful practice, with each breath, of simply observing and noting the lower leg and knee.

As you breathe in again, direct this breath to the thighs and the large muscles under the leg. Notice the weight of the clothing on your body and where these muscles are supported by the surface beneath you. Of course, our legs are held in place by the hips and the bones of the pelvis. Much of our daily work and exercise is in the large muscles of the buttocks. If you find your mind wandering by following an unrelated thought, like that it's silly to focus on the buttocks, or even that the word itself is silly, use that as a cue to mindfully return to your study of these regions and practice bringing your awareness back to the sensation of resting, the weight of your clothes, and each breath that directs you into this part of the body scan meditation.

As you breathe in again, notice the core of your body, especially the solar plexus and third chakra region. Center your focus on the belly, the stomach, and the organs inside. Simply let them be, with each breath, as they are. Observe all of the sensations and feelings, choosing not to follow those thoughts and just participating in your awareness of the body, without judgment.

Back and Chest

As you breathe in mindfully, be aware of the air entering your lungs, filling the lungs, and the inhale becoming an exhale. Scan the upper body, feeling the muscles used in each breath, and the feeling of the ribs expanding and contracting. Each breath occurs in each moment, and as you scan your body, continue to keep your attention focused in the present moment, gently returning your attention to the body anytime you notice yourself following a thought.

Pay attention to the back, the breath being breathed up and down your spine, and notice an ability to breathe through the top of the spine, deep into the core. Notice the sensation of openness with each breath.

Right Arm

Shift your awareness to your right arm, hand, and even to the fingers of the right hand. Direct the next breath as if it is drawing energy from the world around you, through the fingers and hand, and into your right arm. Notice the right arm and its weight as it relaxes with each breath, and the feeling of relaxed and heavy fingers.

The hand is, of course, attached to the wrist. So, follow each new breath up the right arm through the forearm. Notice the elbow and up into the bicep. Notice the awareness of following it up through the right shoulder. Continue breathing in and out, breathing your breath from the tips of the fingers, through the hand and wrist, and all the way up through the arm and shoulders.

Left Arm

Shift your awareness to your left arm, hand, and even to the fingers of the left hand. Direct the next breath as if it is drawing energy from the world around you, through the fingers and hand, and into your left arm. Notice the left arm and its weight as it relaxes with each breath, and the feeling of relaxed and heavy fingers.

Follow each new breath up the arm and through the forearm. Notice the elbow and up into the bicep. Notice the awareness of following it up through the left shoulder. Continue breathing in and out, breathing your breath from the tips of the fingers, through the hand and wrist, and all the way up through the arm and shoulders.

Head and Face

As you breathe the next breath, shift your awareness to the air as it is inhaled. Focus on the point where the air enters the nostrils, noticing the pathway of the breath through the throat and how the exhale feels on the skin of your face. Scan all the tiny muscles of the face as you let them relax. Let the tiny muscles of the eyelids relax with each breath and let the jaw muscles loosen, noting the feeling of deep relaxation in the face. As you breathe in again, let the muscles behind the ears and the back of the head relax, noting the very top of your head and even the highest hair on the top of your crown.

Now notice the feeling of warmth, energy, and awareness that you have in this moment. Notice that each breath is its own time, and that you have created awareness of the body by studying each place and breathing breath into each cell of the body. This is a healing process, in both mind and body.

Suggestive therapy

The ability of nature to heal is always amazing to a child. It seems that every child who visits the zoo is mesmerized by the snake shedding its skin or the wounded lizard that is able to regenerate a part of the tail. Plants also have a healing ability. This is seen in the leaves that are shed and are replaced by new, green, healthy leaves, or by the plant that seems to quickly come back to life with a little water or sunlight.

All of nature is programmed to live, regenerate, and to heal. The same child, who was drawn to the ability of an animal to shed its skin, is also usually captivated by the scabbing of a wound, the healing that takes place underneath it, and the new life that emerges where pain used to be and where injury once was. When you were a child, did any of these things ever capture your attention? Of course they did. What is amazing is that as children, we simply accepted these remarkable abilities in nature, without wondering if they were possible or if it always worked.

There is nothing mystical or magical about this. These are abilities that are natural to the body that we were created with. In paying attention to this, we realize that the body is doing its job automatically by bringing healing energy, light, sound, and feeling to every cell. It does this automatically and every day, even though we usually pay no attention to it, because we usually do have not an injury or wound to draw our attention to it. But it is there, isn't it?

Follow that feeling, that light, or even the voice of healing though your body as blood pumps life into each cell and as oxygen delivers its healing breath to each cell. Focus on that part of the body where you need healing most, realizing that nature always does it job and that today, the recovery process is already underway. With each breath, in every way, you get better and better.

Now, imagine that spot that needs healing is like a child's scab, giving way to life underneath it and realizing that soon the dynamic power of nature will result in new life, where old pain used to be. Each and every day, take a moment to breathe and focus on that breath. Focus on the thought that in every way, each and every day, you are getting better and better. Say to yourself, "Every day and in every way, I am getting better and better. Every day and in every way, I am getting better and better."

Law of Attraction

Pre-talk

This session has been specifically created to unlock the master key to your success with the Law of Attraction. Many people ask me how hypnosis can actually help to unlock the abundance of life and the power of the law of attraction. First, the law of attraction is based on the simple principle that "like attracts like." Charles Hannel, an influential teacher of the law of attraction, has said, "You cannot entertain weak, harmful, negative thoughts ten hours a day and expect to bring about beautiful, strong, and harmonious conditions by ten minutes of strong, positive, and creative thought."

Hypnosis commits this process of strong, positive, and creative thought to the permanency of the subconscious mind. It literally allows us to live abundantly and positively in the soul of our subconscious mind, twenty-four hours a day. By practicing the right kind of hypnosis, it creates a state that is not dissimilar to a magnet where we attract that which is positive throughout the day and night, at a subconscious level.

Secondly, hypnosis is a learning process. It is a skill similar to meditation that charges the batteries of our soul and gives us clarity to discover our true potential. It is like a tool for finding gold among the other minerals of the earth, and will empower you to find abundance in every aspect of life. Would you like abundance in health? Would you like to attract wealth? Do you want to both be able to love others, and to attract other people who truly love you? That is what the processes in this session are for.

Thirdly, I want to point out that the Law of Attraction requires positive activation and this session gives you affirmations in hypnotic trance that use powerful "I am" statements. The Law of Attraction is a way of life, and by participating in this session, you will find the abundance that is most valuable to you.

Induction

Take a deep breath, inhaling excellence and exhaling anything known or unknown that has been a negative component of your day. Begin with the affirmation that states, "I am the life I desire. I am the life I desire. I am the life I desire."

You will find it easy and natural, as you are guided by my words, to deeply relax and access that place in your mind where an inner creativity unlocks your full potential. Do not worry if you need to adjust for comfort. It's perfectly okay to move and get comfortable at any time during this session. Also, do not worry if the mid drifts, dreams, or wanders from each and every word. It is okay to be either very attentive, paying close attention to the suggestions, or to let yourself access deep trance states, choosing to experience this process rather than listening to it.

Scan your body, relaxing any muscles that are holding the tension of the day. Notice how good it feels to let go and allow the muscles to be heavy, relaxed, or weightless, as you take this well-deserved time to create, learn, and activate abundance in this moment. You have already done a great job, following this guidance. As you relax the brow, unclench the jaw, let your neck relax, and your entire body become peaceful, notice something. Without effort, the heartrate has become smooth and rhythmic and the breath has become calm and regular.

This is really lesson number one. By choosing to relax the body, you have manifested and attracted serenity. Some people spend an entire lifetime pursuing serenity, but we invite that which we think. Your action of employing simple relaxation techniques has attracted serenity from the heart, from the breath of life, and into your consciousness as a perfect creation. Remarkable isn't it, how easily the law can be activated?

As you relax, go deeper into the serenity of this moment, knowing that as you listen to my words, the subconscious mind is now being reprogrammed with the affirmations you are now receiving.

I am powerful.

I am good.

I am abundant.

I am the life that I desire.

Notice that these are present tense affirmations, which you are now programming into the mind through this learning exercise of hypnosis. They are powerful affirmations that activate the attraction of abundance. So, as you let yourself go deeper into awareness, creativity, and relaxing hypnosis, recognize the power of now to create the best life of abundance. Now, magnify the awareness of relaxation in the body, letting each muscle from head to toe deeply relax.

Progressive Muscle Relaxation

Add your favorite progressive muscle relaxation here.

Suggestive Therapy

Up to this point, we have focused on relaxing the body. In doing so, you have attracted physical serenity. Moving this to a mental or even a spiritual level is just as easy. It is not a mystery that we can achieve serenity in our spirit. The law of attractions simply tells us to invite this abundance into every aspect of our life and we will begin to feel a perfect peace, a new serenity, and the promises of abundance. Say to yourself, "I am serene in my spirit. I am serene in my spirit. I am serene in my spirit."

Notice that in the time you have taken so far, you have allowed yourself to enter a deep trance state. You know you could move, get up, or end this session if you wanted to. However, it feels so good to be that which you know you are and to dedicate this time to attracting abundance, that it becomes very easy to go deeper into hypnosis and creativity. This is the beauty of hypnotic meditation. It becomes easier to attract success and to activate the law of attraction than it does to hold onto any negativity or continue to believe self-defeating self-talk that you might have previously held on to.

Now, release that which is either known or unknown, by letting any remaining tension or negative thoughts slip from your relaxed body and serene mind. Just let those things go through the soles of your feet and into the floor below you, deep into the core of the earth. You can smile, knowing that this is the point you have taken action to activate the law of attraction in a new, subconscious, and permanent way.

I am going to count backwards from seven to one. In between each number that I speak, I will use a powerful "I am" affirmation. As I count and state the affirmation, repeat it back in your own mind. With each number, each breath, and each affirmation, bring yourself to a deeper place of hypnosis, all the way down to the subconscious level of awareness that is limitless in its ability to create and experience attraction and abundance in every area of life. One . . . I am the life that I desire. Two . . . I am abundant. Three . . . I am perfectly gifted. Four . . . I am harmonious and positive. Five . . . I am creative. Six . . . I am loved. Seven . . . I am thought energy.

One of my favorite books is a simple one that was written by R.J. Banks called *The Power of I AM and the Law of Attraction*. In this book, Mr. Banks tells us that whether one believes in the law of attraction or not, it is constantly present and working in all of our lives. He points out that we often say "birds of a feather flock together," usually denoting some type of ill-advised relationship, or "be careful what you wish for," as some sort of common karmic observation. He writes, "I find it amusing that our ego can acknowledge the law of attraction in the negative or undesirable context, but often struggles with trusting its capabilities in a positive perspective."

This is where we are now. You have created a powerful "I am" trance state by participating in this session and in this moment and in this breath, recognize a new congruency between your negative faith in the law of attraction and a new realization that will guide all future thoughts. That realization is that the law of attraction works in the positive, as well as in the negative.

So, you can adapt this new congruence, now looking at the glass as half full and knowing that you have a powerful ability to speak into existence anything that is important to you, because words have great power. It is from our thoughts that our words create, and your attention to this will always help you to manifest the law of attraction in your daily life. Five . . . four . . . three . . . two . . . one . . .

As you manifest the law of attraction in every area of your life, know that today is the turning point from hope to knowledge, having moved from thinking about manifesting wealth, health, love, and well-being, to taking action and positively creating a new energy. Thought is energy and that energy is creative and is now at work in your life, building a powerful new manner of living, attracting in each and every way that which is positive, creative, and energetic.

At this point you are now ready to step into the three parts of the law of attraction. Those parts are called "ask, believe, and receive." Send out to the universe that which you seek. It has been said, "Ask and you shall receive." Take this time in your own mind to ask the universe for that which you desire most. Visualize that which you want most and claim that by saying to yourself, "I am co-creating with the universe all that is intended for me. I am co-creating with the universe all that is intended for me."

See yourself asking the universe by visualizing your request, giving it to the universe, and being heard by these forces. We know that these forces of universal law are always hearing our words because these promises have been expressed in every major philosophy, by countless advocates, and are imbedded in the hearts of people since the beginning of time itself.

Believe. See yourself stepping into your new abundance, your new love, your new wealth and success. Whatever the mind can conceive, the mind can achieve. This is an axiom that has stood the test of time, so amplify your confidence and your belief, knowing that the law of attraction is at work in all of our lives and your life, because it must work.

How does one receive that which has been asked for? By opening the mind, the spirit, and by cultivating gratitude for the abundance that is available, you have created something wonderful. Gratitude is the attitude that opens the heart, the mind, and the body to healing, well-being, wealth, success, and positive energy. Say to yourself, "I am complete creation. I am a remarkable human. I am that which I ask for and that which I deserve."

Awakening

Use the awakening of your choice.

Fear of Needles at the Dentist or Doctor

Pre-talk

Many people ask me if it is possible to become comfortable with procedures like cleanings, fillings, and the application of a local anesthesia in just a short time by using hypnosis. The answer is yes. The reason is simple. You have opened your mind to the possibility of creating calm and comfort, and we become the thoughts that we think.

In this short time, I am going to teach you three things. First, that you can control your physical responses, creating comfort during procedures, even where you anticipate discomfort. Second, that you can create calmness, not by giving up anxiety or fear, but by simply accepting anxiety or fear. And third, that you can accomplish anything, in any setting, by using the principles this session contains.

Induction

To begin, close the eyes and adjust for comfort. During this session, it is perfectly ok to adjust at any time by altering your posture, your muscles, or by clearing the throat in any way, allowing yourself to become more comfortable. This is lesson number one. You have the ability to decrease discomfort and increase comfort at any time and in any way. If you can do this now, you can easily do this even during a dental procedure.

As you relax, bring your attention to the breath. Become an observer of the breath and pay attention to the way it feels to breathe air in and out. Do you notice the place in the lungs where the inhale turns around and becomes an exhale? Do you notice the temperature of the air as you breathe in and out? Can you study the way each breath feels?

The reason you are focusing on the breath to overcome a fear of needles is simple. As long as you are breathing, you are actually okay, no matter what else is going on at any particular moment. By paying attention to the breath, you are setting aside past experiences and staying in the moment, not paying attention to any anxiety caused by fear. Notice how just by closing the eyes, adjusting for comfort, and paying attention to the breath, the breath has become smooth and rhythmic and the heartrate has become calm and regular. It is effortless.

Now, here is a fact about needles. Today's needles are so narrow and small, it is almost as if they do not pierce the skin, but glide between the cell walls to deliver the relief you want while the dentist works on your teeth. While we all know that being at the dentist is not the most comfortable thing we can do, by focusing on each breath when we notice anxiety, rather than following that anxiety, you will notice an instant sense of calmness, a decrease in heartrate, and increased comfort.

Lesson number two is also quite simple. Panic, anxiety, and fear are only a problem if we follow those thoughts. In other words, instead of using the mind to explore the meaning of these things, or even ways to get rid of these feelings, you can use the mind to label them and accept them. So, during this time, during your actual procedure, or anytime you have a panic, fear, or an anxiety, instead of exploring those emotions and following them with the mind, simply label them. You can simply say to yourself, "That was a fear." or "That was a panic." In those moments, just let it be what it is, using that recognition as your cue to return your attention to the breath. We know that in each breath is a new moment and no matter what else is going on, as long as we are breathing, we are actually okay.

You can actually take a moment now to congratulate yourself. You have come this far and you have learned something new today. You have learned that instead of fearing a process or a procedure, you can be comfortable, calm, and relaxed, in each moment. You can even anticipate having your dental work completed and look back at all you did to apply these new skills, feeling proud of your accomplishment in overcoming any fear by practicing mindfulness and creating comfort.

The upside to this is that from this point forward, any needle from any doctor or nurse is something you will be able to handle. This isn't because you gave up your fear, but because you have not followed it. Instead, you simply return your attention to the breath and staying in each moment. So now, with a deep breath and a smile, open the eyes, feeling ready for the next chapter of your life and ready to complete the tasks of today with confidence, calmness, and new perspectives.

The No Texting and Driving: Enjoy Driving Without Distraction

Pre-talk

People ask me, "Can hypnosis really help a person to stop texting and driving?" While it may seem remarkable, it is not. Distracted driving often happens due to texting, but can also happen by looking at email, reading webpages, or scrolling through music and other files. This is simply a habit. Just like other habits, they can be changed. Distraction with electronic devices is also a function of boredom, and self-hypnosis can equip your mind with alternatives to the mundane travels we often have to experience. Of course, the devices themselves are to blame too. They function as a virtual "tap on the shoulder" and our psychology is just not wired to ignore such an important cue. By learning self-hypnosis, we can break the habit of checking our phones when driving, adapt new perceptual positions on boredom and monotony when driving, and break the internal desire to "just check" something.

Hypnosis is not a magical or mystical state, nor is it a state that is difficult to achieve. Unlike Hollywood portrayals, it really is just a time that you set aside to learn something new that can be useful to you and to commit to a plan of action that you will find helpful. Hypnosis usually uses relaxation as a technique, because in a relaxed state, we can set aside distractions of the day (like our phones), as well as worries about tomorrow, and really just be mindful of the moment. In this resource state, we can access the powerful motivations that set aside dangerous behavior like distracted driving and really make a choice to change. In relaxation, I can also teach you new skills. You might even call these skills psychological tricks because they can reframe how much attention you really give your phone while driving.

Induction

Start by paying attention to your breathing. We usually just breathe without thinking about it, but now pay attention to your breath. Pay attention to the way it feels to breathe the air in, and even follow the inhale. Follow it with your awareness, as it travels through the back of the throat and into the lungs. You are bringing your attention to your breath and even noting that point when an inhale turns around and becomes an exhale. You don't have to speed up or slow down your breath because we are not doing any special kind of breathing here. All you are doing is paying attention to the breath, really observing it, and focusing your attention on it.

As you do this, it is perfectly okay to adjust for comfort and to also notice any muscles in the body where you might be holding onto tension. Just release that tension. You can relax the brow, unclench the teeth, and even drop the shoulders, relaxing the body as you pay attention to your breathing.

At this point, you might notice the mind wandering, thinking, or even wondering about this process. That is perfectly okay. If you are like many of those who are tied to electronic devices, your mind has many things to think about during the day and you are probably really good at thinking about many things at once. You don't have to worry about your mind thinking as you pay attention to your breath, because thinking is just what minds do.

Right now, just relax your body a bit more. Let the muscles that you released tension from become heavy and relaxed, and continue to pay attention to your breath. With each breath and each moment, relax a bit further into a resource state of open-mindedness and new learning. Over the next several minutes, you will be learning new ways to enjoy driving without distraction and to set aside the many burdens of your attention that electronic devices demanded from you. In fact, it feels pretty good to take this time free from distraction to focus your attention on something as simple as the breath. It also feels good to take a few minutes for yourself, doesn't it?

With each breath and each moment, create an awareness of the internal power you have to take control over priorities, demands of others, and even the desire of your own mind to access information at the wrong time. Do you sense that internal power, with each breath and each moment? Have you noticed that effortlessly in this process, your heartrate has slowed and your breathing has become more smooth and rhythmic? That is the power of self-hypnosis. Although I have guided you, it is you that has created this resource state that we call hypnosis and you can congratulate yourself. By taking the short amount of time you have in coming to this session, you have already magnified that internal power of yours.

Deepener

I am going to count backwards from ten to one. With each number that I count and with each breath, let the demands of the day go a bit more. Value each moment, each breath, and your ability to focus your attention. Ten . . . nine . . . eight . . . seven . . . six . . . five . . . four . . . three . . . two . . . one.

To this point, you have learned a process for relaxing by intentionally letting the muscles release tension and you have learned to be mindful of the moment, by bringing your attention to something as simple as the breath. Do you notice how the many thoughts you may have been consumed with at the onset of this session give way to an internal power to accomplish what is important to you?

Metaphor and Indirect Suggestions

In many ways, hypnosis is not unlike untying the knots in an old shoelace. Even when the knots are tough, we know that with the right lighting, the right attention, and perhaps a tool such as paperclip or small knife to create an opening between the tight laces, they will begin to unravel and eventually become loose and usable. For many who wrestle with distracted driving, it is the knots of our job, our kids, our friends, or even social media posts that wrap us tightly in the desire to check every few minutes while driving down the road. But have you ever untied a really tough shoelace? Of course, untying something can lead to things of incredible value. Have you ever received a gift that was carefully wrapped and even tied with a bow? By untying that bow and opening the paper, did your surprise and joy at what was inside make you feel happy? I am really happy that you have decided to give yourself a gift today of learning something new and giving the gift of life to your family, friends, and other drivers on the road.

Transitional Deepener

Five . . . four . . . three . . . two . . . one . . . zero.

Direct Suggestion

Of course, you used to drive without texting. Remember that? It may have been when you first learned how to drive or before these distractions were widely available. We always know we can do something if is something we have done before. So, you know you can set aside the distractions and even do some of the things that you used to do when driving, like sing along with the radio, or mentally planning out your day. In fact, people used to drive just for the opportunity to chat in their head.

Driving is a great chance to be present, if you let it be. You have taken the time to learn hypnosis, not just because it has helped you to create a state of relaxation and internal power, but also to specifically stop becoming distracted by your electronic devices while driving. The good news is that you have already done what is needed to stop texting, reading email, or looking at social media while driving. You have done this by learning to focus on the moment and by paying attention to the breath.

Over the next day or two, the next week, and even over the next few months, you will notice something when that virtual tap on the shoulder is heard. What you will notice is that how by leaving the phone where is, you can return your focus to the moment by simply paying attention to the breath. Although it may sound simple, using your breath as a focal point anytime you find yourself distracted is a powerful way to live in the moment and to let wait what can wait. Anytime you find yourself anxious about waiting, you have another opportunity to intentionally bring your attention back to the breath. Each time you do so, you find it easier to make that simple action a powerful new habit when driving.

Of course, there are also practical things you can do to help you eliminate distraction. Some of them are proactive. By turning your phone to airplane mode, or simply silencing it before getting in a car, the old axiom "out of sight out of mind" comes into play. I recommend always putting your phone inside a pocket or in a car door on silent mode so that the light from the screen will not capture your attention. Of course these are suggestions, but they are suggestions that you can act on with confidence because you have taken the time to really create internal power and change your previous behaviors by participating in this hypnosis session.

You have entered into a new chapter in life, with new patterns and new habits. It is a chapter of life where driving itself is no longer monotonous or boring. Instead, it is one where driving gives you a chance to be mindful and untied from distraction, bringing you a sense of wellbeing. It is a chance to study the sights you pass, the action of holding your steering wheel, and the art of paying attention.

Just as you have become both relaxed an attentive in this session, you can create these resource states behind the wheel of your car. Now do you notice something? That freedom from stress can have a specific application to ending your technology habit while driving. Furthermore, everything in this session can also apply to any other stressor, in every aspect of life. So, intuitively you can practice the principles of releasing yourself from any knots of stress, both in your body and in your mind, as you practice your new awareness of self-hypnosis and mindfulness in every aspect of life.

Although you will be amazed at how this simple guide has helped you to set aside the distraction of your phone, when you take your next car ride and look back and realize you turned the volume off, set the phone out of sight and simply didn't use it, you will also be amazed at how easy it is to sustain these new habits. Of course, there may be a time when you forget to do this and the beep, buzz, or vibration of your obligations and curiosities might "tap you on the shoulder." However, in that moment, you will have the ability to focus your attention on the road and on your breath and to just breathe, noticing the breath and that a ring can just be a ring, without you continuing to believe your old idea that it needs attention now. You know it can wait.

Right now, tell yourself, "It feels good to just be present in this moment and free from distraction." Go ahead and say it out loud, "It feels good to just be present in this moment and free from distraction." And should a time come where that virtual tap on the shoulder asks for your attention, simply repeat this phrase and return your attention to the breath. You will notice that the first time you chose to do this, a new neural association is created. Each and every time it happens, your new action of waiting until your car is stopped, parked, and turned off before you use your electronic device, becomes more and more natural for you.

Awakening

Choose the awakening that is most suited to your client.

Peak Performance

Induction

Use the induction of your choice.

Deepener

Use the deepener of your choice.

Suggestive Therapy

The great thing about hypnosis is that even though I am giving you suggestions, they are not really suggestions that are coming from me. Instead, these suggestions are really coming from within you because you have asked me to guide you through this process and have selected what is most important to you. So, what you are hearing and experiencing is actually a reflection of your own desires and potential.

From this point forward, it will be easy to move beyond average and into excellence. You are moving into health, abundance, wealth, and even moving forward in the relationships that guide you into success. We know this because you have made a decision today to move beyond average and into excellence.

You have taken the first step by reconciling the conscious desires with your subconscious motivations.

In this relaxed state, realize your dreams in this very moment. What exactly does that entail? It entails using this time to use the creative energy of your mind to move out of any comfort zone and into a resource state of success. It means to take some time to really manifest and attract that which is important to you. We know that like attracts like. This is, of course, called the "law of attraction." Thoughts, images, and affirmations that extend beyond what you have experienced so far and know you will experience in the future, are all within your mind's ability to create right now!

Notice how as you create images of success, wealth measured in money, health, or happiness, these thoughts lead to a feeling of knowing that anything is possible. Right now, what can you commit to in action which will lead to those likes attracting other likes? Is it associating with financial winners? Is it learning from people who are wise with their money, or is it simply committing to a process of saving and compounding that which you already possess? What about health can you commit to right now that will attract health? Is it to release needless weight or to change a negative pattern by choosing healthy options for your success? Also ask yourself what you can do in your relationships to create harmony, happiness, and even success. Moving from this session now and into success . . .

Awakening

Use the awakening of your choice.

Hand Tremors

Note to the hypnotist

This is an actual transcript of a session with a real-life client. Adapt this script to your individual client's needs.

Induction

Allow yourself to experience a deeper state of calm relaxation, with calmness in the mind and in the body. If you are carrying the tension of the day, you can relax those muscles of the body. You can let the tiny muscles of the eyelids relax. You can let the forehead and eyebrows relax and unclench the jaw. If it's more comfortable for you, you can even relax your muscles in the shoulders and neck by letting your chin drop towards your chest a little bit. A lot of people find that if they let their chins drop towards their chests a little bit, it will keep them from feeling tension in the neck and shoulders.

During this session, if at any time you need to move just for your comfort, to scratch to an itch, or otherwise adjust, those things won't disturb you. In fact, they will help you to relax ever further by making you more comfortable. You are doing great so far, deeply relaxing and paying attention to the feeling of relaxation in the muscles. You can do that by noticing the muscles of the arms, like the biceps, forearms, and triceps, and letting those muscles relax. Just let the muscles in your hands relax too. In fact, the muscles in your left hand must be very relaxed. After all, your fingertips are resting and they look calm and relaxed, staying in a comfortable state of stillness.

The right hand, which has the fingers curled, may be uncomfortable. Go ahead and uncurl that hand and just let it rest on the arm of the chair. Notice how there is a difference between the tension of the curled muscles and the relaxation of the open hand. As you breathe in and breathe out, you are not asleep but deeply relaxed. You may pay attention to each and every word that I say, maybe even with the conscious mind wondering how this is helpful, but with the subconscious mind learning new things and creating new experiences that can help you in each and every way, each and every day.

Relax the muscles of the belly and the back, the muscles of the chest, the muscles of the buttocks and thighs, the knees, the calves, the shins, and even the muscles of the feet. Adjusting for comfort is okay and even encouraged in order to relax the muscles in the feet and toes. This is only a basic process of physical relaxation, but you will notice the benefits by taking a minute to relax the muscles in the body. Both of the hands are still, breathing is smooth and rhythmic, and the heart rate is calm and regular.

Focus on the hands now. As they relax, think of the word "heavy." Let those hands just be heavy. Even say to yourself, "My hands are heavy. My hands are heavy." Allow yourself to feel the sensation of heaviness in those relaxed hands. Heaviness didn't come from putting a weight on the hands. Heaviness came from just letting those hands relax and become calm, smooth, and rhythmic like the heart, and regular like the breath. It feels really good to notice those hands are just heavy rather than shaking or filled with the tremors, doesn't it? This feeling is not something that I created, but something that you created by simply saying to yourself, "My hands are heavy. My hands are heavy." You have the power to really create sensations that are important to you.

For example, think of the word "warmth" like the warmth that comes from the sun or like the warmth that comes from the inside of the body. Think of the word and sensation of warmth in the same way that you thought about the heaviness of your hands. Let those hands feel the sensation of warmth. Say to yourself, "My hands are warm. My hands are warm." Let those warm hands notice the sensation of heaviness and heat. You are saying to yourself, "My hands are warm and heavy. My hands are warm and heavy. My hands are warm and heavy."

Our conscious mind knows that at any time should you want to, you could move your hands. However, it feels so good to let those hands relax and become both warm and heavy. You'll find that if you try to lift those hands, they become weighed down. If you try to lift those hands, they become heavier and heavier. In fact, if you try to move those hands, the tremors in the hands will simply become heavier and heavier, locked down even tighter. Now, deeply relax.

You are doing perfect, with each breath, doubling the sensation of relaxation and calm. With each number and each breath, let yourself drift into a slower state of relaxation or hypnosis. Five, four, three, two, one, zero. You are doing perfect. You are warm and heavy, weighed down by the comfort of yourself, with a smooth and rhythmic breath and calm heart rate. You look very relaxed. You are not asleep, but deeply relaxed.

Go ahead and make an "okay" sign with the right hand. Go ahead and do that now. Just touch that thumb and index finger together on that right hand. Touch them, hold that tension for a moment, and then relax your fingertips. Let them relax, noticing the difference between tension and relaxation. Again, touch the fingertips together of the thumb and index fingers, press them hard, and hold them tightly for a moment. Now relax the fingers again, letting those fingers become simply relaxed like they were a few moments ago, noticing the difference between tension and relaxation.

Can you feel that difference? This is what we call an anchor. It is something that you can do any time during the next day or two, or even in the next week, month, or years to come. If you have any fear that you might experience a tremor, simply touch that thumb and index finger together and press for a moment. Then, relax them and notice a sense of heaviness and warmth in the hands, allowing yourself to return exactly to this state that you've created right here and right now.

It is amazing that we have the ability to do that; to take control over our bodies by doing something as simple as touching the thumb and index finger together, relaxing them, and then returning back to this state of relaxation. In fact, with each number and each breath, you can allow yourself to go into an even deeper state of hypnosis, calmness, tranquility, and stillness. Ten, nine, eight, seven, six . . . You are doing perfect. Six, five, four, three . . . Just allow yourself to go down now, completely relaxed. Two, one, zero . . .

Knowing that by coming here today, you have created success, able to easily achieve a state of physical calm and mental learning. You can even feel the power and the ability to shoot guns, reach for your credit card, or eat a meal. Feel that state of calm in the little muscles of the fingers, the tiny muscles of the hands, and the little muscles of the wrists. Right here and right now, while your hands rest and remain calm, look and see that those hands are enormously still, heavy, warm, and calm, just like the breath and just like the mind.

Remember that this state is not one that I've created for you. This is a state that you have created within yourself, by listening to the instructions and by following the suggestions. Because of this, we know that when you step over the threshold of the door and leave the office today, go down the elevator, or get in your car, that the change has been made. In fact, you will first notice it when you reach for your keys to open your car door and find that it's easy to do that. As you drive home, you will be alert, awake, and oriented. However, you will notice that your hands are feeling calm and you are relaxed through the rest of the night. Even if you look for a tremor that may have been there in the past, you will notice that with each breath and with each heartbeat, your hands are calm, still, confident, and relaxed.

Over the next couple of days, you will notice something else, and that is the color red. It will just be brighter, sharper, and crystal clear. It could be a red QuikTrip sign, a red cup, somebody's red necktie, a red piece of paper, red on the TV, a stop sign, or a tail light. When you see the color red, not only will it appear clearer, it will actually bring a smile to your face. It will be a reminder to you that you've actually accomplished something important in today's session, and your smile is recognition of that.

You will find that by practicing what you have learned here today, the tremors that have distressed you to this point are nothing to worry about. Because you've learned this new technique, you can dial back or turn down the volume or intensity of those sensations. So, you have to congratulate yourself for learning something new and for taking control over an area of your life that is so important for you.

And as you breathe in and breathe out, really notice the oxygen filling your lungs and pay attention to what it's like to just breathe. Follow the breath as it travels through the nose and into the lungs, and what it feels like to exhale gently. Every time you breathe, of course, oxygen goes through the lungs to every cell of the body. Over the next couple of days, weeks, and even months, you will find a happy sense of calm overtake you in this relaxing way. You will find it easy to go to sleep once you get up in the night to go to the restroom. It will be easy to drift back to sleep, comfortable and calm, as if you had never even gotten up in the first place, continuing your night with the same level of sleep, calmness, and relaxation.

As you breathe in and breathe out, I'd like you to pay attention to your feet and the heaviness of your feet resting on the floor, your hands resting where they are, and your breath going in and out. Become aware of the spot directly in front of your vision; the place that you looked at right before you closed your eyes. Now, move your attention back to that mental place that is halfway between where you are and where the wall is. You can move your attention all the way back to that spot on the wall where we started, even though your eyes are closed.

When I count to three, you can open your eyes, focusing on the spot on that far wall. You will find that when you open your eyes, you are alert, awake, oriented, and ready for the rest of the day. One . . . pay attention to the breath, congratulating yourself for creating these experiences. Two . . . take a breath, becoming aware of the muscles in your body. Three . . . just open the eyes, paying attention to that spot on the far wall, feeling fantastic and aware of something new and of success. It feels pretty good to go through that process to learn those things, doesn't it?

Hypertension

Note to the hypnotist

The metaphors that are used in this script have been abbreviated from a CD produced by David Parke, called *Prairie Dogs*. These indirect suggestions were designed for anyone dealing with hypertension, but they are also highly adaptable to the specific needs of your client.

Although the goal of this session is to lower blood pressure, medication should never be discontinued or lowered without the approval of the prescribing physician.

Induction

Use the induction of your choice.

Suggestive therapy/Indirect Suggestions

As you continue to breathe in and breathe out, relax completely, all the way down now. Five, four, three, two, one . . .

I want to share a story with you. In Kansas, there's an animal that's called a prairie dog. In fact, many parts of the Midwestern United States have prairie dogs. These animals are really pretty fun to watch. They dig tunnels that stretch for miles underneath the top of the earth. Some of them actually create great cities underneath the ground.

What's interesting is that as a prairie dog burrows through the earth, he removes blockages, dirt, rocks, and perhaps even chews through roots, leaving a perfectly clear tube underneath the surface of the earth. These tunnels can be quite long. In fact, a prairie dog can actually run through those tunnels easily. Sometimes the tunnels connect to other tunnels. The prairie dogs reinforce the corners and clear away anything that would make for an un-smooth transition. It's amazing how in a relatively short period of time, large amounts of earth can be moved by these prairie dogs, clearing away any obstacle that's in their way.

Now, I want to share an observation that I made when I was in New York City. It was in the winter time and from my hotel, I noticed that it was starting to snow. My hotel actually overlooked Central Park and it was really beautiful watching the snow fall. I think the snow must have fallen all night because when I woke early in the morning, the city seemed covered by a white blanket.

As I looked outside of my hotel window, I could see that the city ploughs were busy clearing each and every street. They go up and down the Main Street and they would go in and out of the side streets and the alleys. They paid attention to any area that was blocked, removing the snow early in the morning to make sure that by rush hour, taxis, limos, cars, and motorcycles would be able to freely access all of those roads, highways, and byways in New York City.

Direct Suggestions

These are not things that I think you should do. Instead, these are things you've asked me to suggest to you because you came here today with a desire to address this issue. We know that perhaps the best medication for blood pressure is sweat. So, by increasing your physical activity each day, you'll notice that your body responds to that by decreasing unsafe or distressing levels of blood pressure.

At any time over the next week or two, if you were to find yourself experiencing a high level of emotional or physical stress, just make the decision to take a moment to sit and still the body. Relax the shoulders and even close the eyes. Just return back to the state that you've created here today that we call hypnosis.

Of course, it's important to note your progress. So, each day at the same time and in the same position, monitor your blood pressure. Notice that because of the changes you've implemented and the strategies you've exercised, the number responds each and every day. This decreases your risk of any problems. And because you care for yourself deeply and are committed to this healthy path, you find that making changes such as decreasing your caffeine consumption, avoiding excessive salt, and avoiding tobacco, are all things you find easy to do. It is no longer a struggle because of how much you care for yourself.

Each night after you take your blood pressure, just take a moment to practice the principles of what is call Autogenic Training. We will practice this right now. Simply closing the eyes and relax. Let your hands be heavy and say to yourself, "My hands are heavy. My hands are heavy." You can also say, "My hands are warm. My hands are warm. My hands are warm." Let those hands become warm.

You can also say to yourself, "My forehead is cool. My forehead is cool. My forehead is cool." Now, just let your forehead become cool. In fact, take a minute to simply focus on your forehead and say to yourself, "My forehead is cool," noting the ability you have to sense coolness in your forehead. Again, say to yourself, "My forehead is cool, my forehead is cool." Did you note that sensation? Did you note that feeling of coolness in the forehead?

Of course, we know that if you can create warmth and heaviness, then you can create coolness. If you can create those sensations, you can create clarity in every vessel of the body. You can create smooth and rhythmic breath and smooth and rhythmic blood.

Awakening

Use the awakening of your choice.

Cancer Recovery Support

Note to the hypnotist

I neither endorse nor oppose the choices a client makes with their doctor. This is how one always remains within the scope of practice. Our role is to support the health of the client, within the context of the decisions that they have made with their oncologists. It is also how we build referral sources. For example, if an oncologist was to make a referral to you and you tell the client that they won't need radiation or chemo, not only are you grossly out of the scope of practice, but you will not be receiving anymore referrals either. Medical decisions must always be made between the oncologist and the client, not between the hypnotist and the client. Our job is to support them, no matter what treatment choices they make.

Induction

Use the induction of your choice.

Deepener

Use the deepener of your choice.

Suggestive Therapy

I don't know if you've ever been to the beach and tried to hold sand in your hand. But, when you try to hold sand, it might not stick together well because of either too much or too little water mixed in. The more water it has, the weaker the bond of the sand becomes. Conversely, if there is too little water, it won't be held together either. It's almost as if there's a specific proportion where the sand and the water create the strongest mixture.

Our bodies are like that. The healthy white blood cells are the strongest. They're held together with the exact amount of the correct forces to sustain the life. However, cancer cells are not healthy. They are weak. They are like sand that when you try to hold it in your hand when it has no water mixed in, it easily falls through the cracks of the fingers. Or it may be like that sand that has too much water mixed with it, so it simply drips through the fingers like soup.

Now, focus on that place in the body where the doctor told you the cancer resides. As you focus on that spot, recognize that each tumor is composed of millions of cells. However, these are unhealthy cells. They are not strong cells. They don't have the essential proportions of life- sustaining materials. So, just see them falling apart, unable to group together.

You can also think of your healthy cells like a house of bricks and the unhealthy cells are like a house made of straw. You can just see the wind of treatment blowing through and allowing the strong cells to remain strong and the weak cells to simply fall apart.

As you continue to close your eyes, visualize the treatment that you are receiving as powerful and strong. You've chosen to take certain medications and to have certain processes and procedures done. However you visualize strength, imagine that strength coming into your body and doing what a strong and positive treatment can do. Imagine it zapping, removing, or eliminating those cancer cells, leaving behind exactly what we need to sustain life. In fact, you could even visualize that treatment as a superhero.

Continue to relax. Five, four, three, two, one . . . Imagine the white blood cells being the good guys in white hats. Of course, we know from watching childhood movies that the guys in the white hats always win. So, see those white blood cells as they are. They are the strong and good sustainers of life. See them attacking those cancer cells, always winning each battle and defeating the cancer cells in the black hats.

Each day during this course of treatment, you've paid attention to the changes that you need to make to take care of yourself. Let me congratulate you for making some serious changes in your diet and increasing your water consumption. You can actually visualize that the water that you're drinking is flushing through the body and flushing out cancer cells. Visualize the food that you're eating and the nutrition being dispersed through each cell of the body. Just imagine the nutrition from that healthy food that you've been eating simply eliminating any and all toxins, carcinogens, and even the cells of cancer themselves.

See yourself as you know you will be, three months from now, six months now, and twelve months from now . . . cancer-free. Émile Coué told us that, "Whatever the mind can conceive, the body can achieve." So, see yourself receiving that message from your oncologist that the cancer is gone. Many other people have received that same message, probably even from the same oncologist. So, as you look beyond the present and towards the future, recognize that the "you of next year" is the same you as you are now. Therefore, associate into that step and into letting the "you of the future" become the "you of right now."

Awakening

Use the awakening of your choice.

Wound Healing

Induction

Use the induction of your choice.

Deepener

Use the deepener of your choice.

Suggestive Therapy

In a few minutes, I'm going to give you some direct suggestion that can help you utilize the skill of hypnosis in recovering from your wound. But right now, just imagine being at the sea. Perhaps you are relaxing on the sand or playing in the water. Or, maybe you are simply observing the movement of the waves or people-watching. There is always great people-watching at a popular beach. All day you can see people playing in the sand, building sandcastles, looking for shells, and picnicking. At the end of the day and after all of this activity, the beach gets pretty rough and certainly looks like it has been used.

You can imagine looking down at a view from the sea, from the balcony where you are staying. You can see the high tide and all of that water rolling in. The tide gets higher and higher on the beach. By the middle of the night, you can see the water rolling across the whole beach. Each time the tide comes in and goes out, it carries a little bit of the beach with it.

Just imagine taking a walk on the beach, early the next morning. It is amazing how smooth the sand is and how the natural course of the ocean has simply cleared away the blemishes, debris, and the injury from the previous day's activities. It's amazing how nature has, in almost every aspect of our world, a unique ability to promote healing, health, and wellness.

As you rest there, deeply relaxed and with your eyes closed, bring your mental attention to that place where that wound resides. With your eyes closed, bring your attention to that spot and see that wound as it is, right now. With that creative and adaptive part of the mind, imagine what that place will look like after the period of healing. See that spot as it will be in the future, looking healed, smooth, and safe. Are you able to do that? I'm going to ask you a question. Are you able to see the contrast in the images between the way it is now and the way you know it will be?

Now go back to that image of your wound as it is right now. Certainly you've seen a time lapse video before. I once watched a video that had been taken over the course of a year. It was amazing. It was created by a man who took a picture of himself every day for a year. Then he condensed it down to a one minute video. The time lapse showed things like his hair changing and his beard growing. You've probably seen a time lapse video before, or maybe a video where the sun rises quickly or the sun sets slowly. Have you see that type of video before?

Now, I want you to create a time lapse video or a time lapse image. See that wound as it is now and know how that wound will be when it is healed. Create a time lapse video, really focusing on that spot where that wound is. Just see it on that mental video that you created . . . healing, and then being completely healed.

Direct suggestion

We heal from our wounds not only because of a passive mental process, but also because of specific actions we take that promote healing. Because you care for yourself, it will be easy for you to follow the doctor's instructions for caring for that wound. You find that with each change of dressing, with each washing, and with each aspect of wound care that you engage in, the healing process is encouraged. You get better and better, faster and faster, each and every day.

Of course, the body is comprised of many different macronutrients and micronutrients. Our bodies heal best when protein levels are high. So, in the next couple of days and the next couple of weeks, you'll increase your protein consumption in your meals. You make sure to take care of yourself to promote that healing.

Wounds also respond well to vitamins and minerals. So, choosing foods that are highest in nutritional density is something that you'll want to do over the next couple of days, weeks, or even over the next few years. You make sure you consume enough Vitamin C and choose foods high in zinc, because it gives you the knowledge that your wound will heal faster, more comfortably, and with ease.

Awakening

Use the awakening of your choice.

Seasonal Affective Disorder

Note to the hypnotist

You might wonder how hypnosis can be helpful in SAD treatment. Hypnosis can be utilized in a way to create change in physical awareness and to create hope through visualization. It can reframe the meaning of metaphysical symptoms and can be used to motivate and train new coping behaviors. Finally, it can be used to stimulate the pineal gland.

Induction

An autogenic training induction is a recommended.

Deepener

Use the deepener of your choice.

Suggestive Therapy

In this state we call hypnosis, where you are never asleep, but deeply relaxed, you will be able to accomplish so much today. First, you will be able to learn how to change your body's comfort level. During the induction, you already learned you have the ability to create warmth simply through thought. Second, you will be able to use that creative part of the mind to implement new solutions, where in the past there may have only been frustration with the seasons or even depression. Third, you are going to have the opportunity to step into a new chapter of life where even though seasons might mean change, they no longer have to mean despair. In fact, by simply committing to this session and by following these instructions, you have already begun a new chapter of your life. For that, you can congratulate yourself. In fact, just by finding new solutions to seasonal affective disorder, you have probably noticed a new feeling of hope. You know that from this point forward, by learning new things, the shroud of depression has already been lifted with the warm glimmer of hope being present.

As you relax with each breath, deeper and deeper, become aware of that part of the mind where awareness is created. Deep inside each of our minds is a capacity to create. By simply bringing your awareness into this part of the mind, you have begun to create. Many people find that when they begin to create, it is in the form of visualization. Visualization is simply seeing new things, almost like a vivid dream where you are aware that you are creating. Other people find an inner voice is created, guiding you to a new awareness. Either way is fine because there is not a right or a wrong way for you to experience hypnosis. This is your time and your place.

Imagine yourself stepping into a new chapter of life, from darkness and cold, to that same warmth you created just moments ago. Imagine the radiant sunlight in this new chapter of life, with a smile on your face, as you realize that what the mind can conceive, the body can achieve. Notice how, in just a few seconds, you were able to use that creative part of the mind to feel, if even for a moment, a lifting of darkness and an image of new light. That is awesome, isn't it? Just like you were able to notice or create warmth, you have now used the imagination to create a new chapter of life. The power here is that we have within us the ability to create and change. At any time and in every way, we can gain control over something that once made us feel powerless. Five, four, three, two, one . . .

Years ago, an anonymous poet wrote:

Dead branches
White snow

Blue sky
The sun's glow

Red leaves
The wind's blow

Green Grass
The river's flow

This is what makes the year beautiful
The four season's slow pace

It makes our lives colorful
And gives us the example of grace

Take the time to contemplate
What nature gave us

Because you never know when it'll end
Because of all our mindlessness.

I have read you this poem because it is an essential reminder to us of the four seasons. Even when we feel stuck or trapped in the frustration of seasonal difficulties, absolutely nothing stays the same. As surely as the sun will rise tomorrow, each season will transition to a new season. So as the poet exhorts us, "Take time to contemplate what nature gave us."

I know you have been frustrated by the current season, but within each season is a gift, even when it is hard to see. The poet tells us the seasons change at a slow pace, but they always change. Perhaps the slowness of each season is to give us an opportunity to find that which is beautiful. It is easy to resist the seasons or to dislike and judge the season, but right now and right here, you are safe no matter what the season is. Your heart rate is slow, your breath is smooth and rhythmic, and you are comfortably taking this hypnotic break from your stress. It feels good doesn't it?

This tells us that no matter what the season, you can create acceptance for its presence, neither liking nor disliking that which is outside. Instead, simply focus inside on the part of the mind that creates awareness and acceptance. Just use this time to be still, relaxed, and comfortable. Of course, if you can be comfortable here, you can create this state anywhere and at any time. The ability to experience resource states is one of the great gifts of our creativity. These states can include things like comfort, warmth, hope, and acceptance.

Do you notice the difference in how you feel from when we first began this session, to what you have created now? Did you notice the difference in your stress level, comfort level, and even in your ability to feel hope and happiness? In fact, you can even amplify those feelings of hope and happiness. You have the ability within the creative mind to activate the pineal gland (the third-eye) and to benefit from the hormones that regulate our days and nights and our level of happiness, despite external experiences. Go ahead and try it now. Be aware of your level of hope, happiness, or any other resource state that is important to you.

Now breathe in. As you breathe in, let the breath energize you and amplify that experience of hope or happiness just a bit. Do you notice it? Now breathe in again, taking in another fresh breath and breathing in hope and happiness. Even say to yourself, "I am hopeful. I am happy." Now, amplify it a bit more. It's pretty awesome to be able to amplify those feelings. You might even notice the corners of the mouth twitching and becoming a slight smile as you amplify it even further, creating an awareness of your own ability to create feelings, at any time they are useful to you. Five, four, three, two, one . . .

Direct Suggestions

Over the next day or two, you will notice an ability within to be mindful of your ability to create, to accept, and to embrace the seasons, knowing that nothing ever stays the same. In the days to follow, seek out an opportunity to be in the daylight, knowing the sun is a powerful force for life and even when hidden from view, its radiance is still present.

When you are next at the store or even shopping online, pick up some light bulbs labeled "full-spectrum light" and begin to replace all of the bulbs in your house, when they need replacement, with these new bulbs. Even if the light in your bedroom and bathroom does not yet need replacement, replace those lights now with the full spectrum bulbs. Doing so, you will find that the radiance of these lights changes your body, which changes your feelings. Each day, look for opportunities and time to walk, just for the sake of walking. Increase your daily activity each day, taking more steps today that yesterday, more steps tomorrow than today, and increasing your steps each and every day until you reach 10,000 steps daily.

Awakening

Use the awakening of your choice.

Dozens Of Simple Suggestions For A Variety Of Issues

Smoking

1. Because you have come here today, you know you have already entered a new chapter of life. A smoke free life, where energy is derived from breathing oxygen.

2. You have easily discovered already that any desire to smoke that previously might have existed has been replaced by a desire for health.

3. Like an astronaut discovers the freedom of defying the laws of gravity in a new atmosphere, you have freed yourself from any pull that nicotine has ever had.

4. Discovering each day, that it not only gets easier to be a non-smoker, but that it has always been easier to be a non-smoker. The hard work was labored breathing, cleaning your environment, having to rush for a last minute trip when you discovered you were out of cigarettes, and having other people shun you.

5. As a caterpillar morphs into its full form as a gorgeous butterfly, you now have stepped into a new, beautiful you.

6. Should you ever find yourself with a self-defeating thought, like "I should have a cigarette," you will recognize it immediately and replace that self-defeating thought with the truth: "I am happy, joyous, and free from cigarettes."

7. In the past you may have noticed the clock, which often told you it was time to smoke. But it wasn't really the clock saying it was time for a break, but rather your own mind. We know this because clocks can be changed: we spring forward in April, and fall back in October, and when you see a clock now, it no longer means it's time to smoke, but rather just what the time on the clock says.

8. When you see other people smoking, it will no longer be a cue for you to smoke. It is not something you miss. Rather, it is simply the choice that another person has made, which is a different decision than you.

9. People will notice the gleam in your eye, the glow of your skin, and each morning you will notice this too: reinforcing your continued desire for success.

10. When sitting in that chair or place where you often smoked to relax, you will bring yourself back to this session rather than smoking, practicing the principles of self-hypnosis that you have learned hear today.

11. It's amazing that the transformation has already taken place. It's not that you now need to become a non-smoker,

it's that you are already a non-smoker. Breathe in now: notice you are not smoking, you're just breathing. You are a non-smoker

12. As the sands in the hourglass sift, you have moved as well. From one side to another, you have changed with meaning and intent.

Weight Loss

1. When our session is done today, the first thing you will do is download a pedometer app on your phone, even before you leave here today. You will use it to measure your steps, taking more steps today than yesterday, more steps tomorrow than today, and increasing your steps to 10,000 steps per day.

2. Any time you are in a restaurant and order a meal, your new pattern will be to ask for a go box when your order is delivered, immediately cutting your portions in half and taking half of that meal home for tomorrow's meal.

3. Eating is now a meditation for you, where you will mindfully enjoy the flavours, textures, and tastes of food, slowing down your eating taking twice as long to eat any meal as you have in the past.

4. For each meal you prepare, your first thought will be: "How can I eat more of the nutrient-dense food?" saving that which is least nutritious for the smallest portion.

5. When at a buffet, a potluck, or a shared-plate dinner, you will no longer see it as an opportunity to eat all you can, but instead as an opportunity to make two or three healthy meal choices in the correct proportions.

6. It will be intuitive to listen to your body, knowing exactly how much to eat, and finding it easy to stop when your body has told you its needs have been met.

7. Everything in moderation is the new rule. Which doesn't mean you can't have indulgences, but simply chose moderation as the first option.

8. From this point forward, fried food, snack food, and other low-nutrition food will become your last choice rather than your first choice.

9. By using a calorie-counting guide, you will track your daily intake, and match that against your daily expenditure as measured by your FitBit or health app.

10. When noticing a sensation of hunger, rather than trying to make it go away you will choose to pay attention to it, recognizing that it is not bad. It is just a part of the digestive process, and that hunger tells us our bodies are working, not that it is necessarily time to eat.

11. See yourself now as you know you will be: a week from now, or ten days from now, two or three or four pounds lighter. And a month from now, eight or nine or ten pounds lighter. And even a year from now, over 100 pounds lighter, knowing that what the mind can conceive, the body can achieve.

12. Old habits like looking for the closest parking spot, riding the elevator, or even driving a short distance, are being replaced with new ideas and new behaviors that increase your daily activity. You will relish any opportunity to chose the furthest parking spot, knowing that you not only get more steps out of it, but that your car will have less door dings.

Academic Performance

1. Notice how relaxed you feel right now. When entering a test taking room, stop and pause before you take your seat. Take in a breath, and with the eyes either open or closed, bring yourself back to this calm, relaxed state you created right here right now.

2. As soon as you sit down to take a test, particularly a timed test, before you even begin, flip to the last page and imagine what it will be like to answer all of the questions and finish in time.

3. When taking a test, and hearing the answers in your mind, trust your intuition rather than second-guessing, and simply move forward with confidence.

4. When taking a test, should you notice a tense grip on your pencil, or a knot in your back, or clenched teeth, simply scan your entire body for tension, breathe, and relax any muscles that need to be relaxed, moving forward calmly and easily.

5. And if you notice any sounds from inside or outside of the room, or the movement of people that you might interpret as a distraction, simply use them as an indicator to reinforce that you are exactly where you should be, doing exactly what you should be doing, and are no longer bothered by said distraction.

6. Without stressing over your ability to recall what you have memorized, simply visualize that which you know on a chalkboard in your mind, reading the answers you remember, and writing them confidently.

7. Because now you have made a decision to improve your performance, it is as if you have always found it easy to succeed.

8. By setting aside a time and place to complete your homework, and choosing to be in that time and place until your work is done, you'll find it's actually faster than procrastinating and that the rewards of timely completion free you to do everything you enjoy.

9. We know that each academic success brings us a greater level of financial success, and so even if your motivation is only the increased rewards of your studies, continue forward, because wealth is the result.

10. Rather than saying to yourself "This is not something I understand, or something I cannot do," use a daily affirmation in times of stress, such as: "I am mastering new things, I am mastering new things, I am mastering new things." And challenge yourself for success.

11. And so right now, look inside of yourself and find that which is valuable in your studies congruent with your goals. Embrace that congruence, and recognize that you are reaching each and every goal in each and every class.

12. Imagine another you, outside of you, watching you, observing you. Float outside of yourself right now, and observe the other you from a distance. See yourself studying, mastering, passing, and completing.

Wealth Creation

1. You will find that by using daily affirmations, the financial success you experience will be like opening a treasure chest of gold.

2. We know of course that it is never too late to succeed. Wayne Dyer had never written a book at age 55, Colonel Sanders was living in his car at age 55, and Grandma Moses hadn't painted a painting until she was 78 years old. So set aside any thought in your mind that it is too late to succeed, and embrace the ability to generate wealth.

3. It is important to ask yourself and the universe for your success. It is okay to believe that you are entitled to wealth, and taking action, receiving those riches is a formula that has proven itself time and time again.

4. When Napoleon Hill wrote the book "Think and Grow Rich," he created a classic which has been the blueprint for many others. And so, by using your power of your

mind to think and grow rich, we know the outcome will be the same for you as well.

5. The Law of Action underlies the Law of Attraction, and so each and every day as you take each and every action, ask yourself: "Does this bring me closer to, or further from my goal?" Choosing to increase the actions which generate money, and decrease the actions of procrastination.

6. Each and every day, ask yourself "In what new way can I create, act, and achieve?"

7. In every way I'll begin creating new wealth by gratitude for the wealth I already possess, not looking at what I lack but finding instead what I have.

8. The Law of Vibration tells us that everything is in motion, that thought is energy, and that where my thoughts go is where the energy flows. And so I will think thoughts that move me in the direction I wish to go.

9. Benjamin Franklin said, "A penny saved is a penny earned." And so rather than cutting out what is important to you, make financial changes by saving for that which is important to you, seeing the action of thrift as a value of wealth.

10. No longer being fearful of risk, be willing to take action on your ideas, recognizing that there is no such thing as failure, only feedback, and that each calculated risk produces rewards, either in cash or in learning. Both lead to wealth.

11. Give up now any notion that wealth is evil or bad or sinful, recognizing that great wealth has the ability to create great changes, not only in ourselves but in our families, communities, and the world.

12. You are not only gaining abundance, but an ability to tap into the resources of the subconscious mind, that part of you that knows you are not only driven to success, but entitled to success, and can claim abundance.

Session Optimizers

1. You have done great work in this session. These suggestions are not suggestions that come from me, but rather are suggestions that you've asked me to make, that you can claim, that you can own and move forward from here into success.

2. By coming here today, you have chosen to make a change, but that transformation has already taken place. It is not something yet to be done, but something you have already done simply by coming here today.

3. And notice the resource state you have created. This is not something I have done to you, but something you have created by following my instructions, which means that you can take everything of value in this session to every part of your life at any time or any place, with or without formal hypnosis.

4. And you can congratulate yourself: You have done a great job today, easily discovering the pathway to success that clearly will work for you.

5. And so breathe: noticing that in this moment, any doubts of the past or fears of the future are not present because in this moment, marked by this breathe, you are doing exactly what you need to be doing.

6. Notice how effortlessly and easily you were able to go into trance: if you can create a resource state of trance, you have the ability to create any resource state that is of value to you at any time.

7. Now knowing what resources you can access. You will have the confidence to act in alignment with the values and desires, knowing that success doesn't come from the outside, the external world, but from the inside, the world within.

8. In each and every way, each and every day, I get better and better.

9. Listen now to your internal dialogue: It's telling you "I can." It's telling you "I am." It's encouraging you to succeed.

10. And take a moment here in silence to release anything, either known or unknown, that you need to release, either high in the sky or into the core of the earth. And do that now (adapted from John Cerbone).

11. For every motion, there is an equal and opposite motion. And while we know this is true in the physical world, it is

also true in the thought world. And so as you move away from that which has been self-defeating, recognize that you are moving towards that which is perfect.

12. At one time the tallest building was the Empire State Building, and at one time the tallest building was the Sears Tower. And now the world's tallest building is the Burj Khalifa. Someday, we will build an even taller building, which tells us that each session builds on the success and learning of the previous session, and that infinite possibilities exist for you to rise tall in every situation.

Anger, Anxiety, And Depression

1. When angry recognize that it is neither good nor bad, it is just energy, and that you can use and direct that energy into any choice that is good and for you.

2. Notice that any time you are angry, you are also something else too. Is that fear? Or hurt? Or pride? Realize that is not anger that is the problem, but rather the core emotion which you can resolve.

3. And while anger might get the job done, ask yourself now, "What other ways could I accomplish the job?"

4. Anger is often created by looking outward at what needs to change in others, but in this time and place the only one who is here is you. And so the first question really is, "What needs to be changed in myself, and how can I make that change?"

5. Depression can be a signal that we need to slow down, take a break, step away from a situation. And so in this place and this time, take that break. You're doing perfect. But move forward form this session with the energy you've gained from taking this break.

6. As you visualize depression, is the image hazy and fuzzy? Are the colors dark and muted? Is the soundtrack slow and somber? What if you change the music of depression, letting yourself be depressed but hearing upbeat music instead, and changing the colors in your mind to all of the colors of the rainbow. What happens then? Check. Does depression feel any different? I thought it would.

7. Depression is not something you are, but something you notice. What else do you notice? The feeling of your body relaxed in the chair, a tingling feeling in your skin, maybe even a part of you that has some joy somewhere. It was easy for you to notice depression, but what else can you notice? Can you notice anything else coexisting with that depression? And now amplify that awareness, increase the volume of that awareness, and let it begin to become stronger than the depression.

8. If you have the ability to be depressed, if you have the ability to notice depression, you have the ability to notice what's missing. Explore in your own mind options to fill that void, actions you can take, places you can go, people you can affiliate with. And when our session is done, in every way and every day, seek that which is filling, rather than empty.

9. Notice how by practicing hypnosis the heartrate has become smooth and rhythmic, the breath become calm and regular, and that this is the opposite of anxiety or panic or fear. And so at any time and at any place, if you're distressed by anxiety, practice these principles of self-hypnosis.

10. Any time you feel anxiety recognize that that anxiety is just a thought, and that we are not just our thoughts. And so rather than following that thought and building that anxiety, just recognize anxiety is a cue to bring yourself mindfully back to the moment, and to be present without following a thought.

11. It's amazing how the research shows that most of what we worry about never materializes. And knowing that, means you can forget about your anxieties, focusing on the present.

12. There is of course no reason to dwell on the past, as it is history. And no reason to fear to future, as it is not here. The only moment we have is now, the present, and that is why it a gift. For no matter what has gone before, or what will come, right now, everything is perfect and exactly as it should be.

Sports Performance

1. As easily as you can focus in my watch, a far point on the wall, or anywhere else your eyes are fixated right now, you can fixate your attention in the sweet spot, that place where the perfect shot will land.

2. As easily as you can create relaxation in this session, you can use these same skills to reduce stress prior to a big game, easing into it with confidence, power and skill.

3. You will discover when this session is over, a new confidence in every area of life, especially in your competitions with others in your sport. Confidence is always self-generated, and what you have generated here can be generated anywhere an in any situation.

4. Focus, energy, speed, endurance, these are all resource states for the athlete, but resource state you can step into right here right now. Be endurant. Be fast. Picture it in your mind or feel it in your soul. Be focused, energetic, knowing that what the mind can conceive right here, is easy to make a reality on the game field there.

5. There is no fear in success, there is only confidence that win or lose, it is how you play the game. The way to stop judging yourself, is to use each experience as a learning endeavor, realizing that there is no such thing as failure, only feedback.

6. You know of course that n like, there are many things that are better accomplished by a group. One ban can build a house, but ten men can build it faster. As you play on your team, recognize that the fastest past to victory for you- is being part of a team with all, and sharing success by each using their individual strengths.

7. In the past you may have experienced injury, or a scary near miss, or something that has held you back from all

the success you know you can crate. After all, you told me that if this hadn't happened, then this would be ok. But yesterday is history, the past is gone, your injury has healed and your doctor has cleared you – and so when leaving this session, and heading into the game, it is a new opportunity for performance, completely independent from any past experiences.

8. Aa a child who loves to play a sport, becomes a master of the craft, but now must also take care of the business of sport, these new realities are no longer contradictory, but rather two sides of top career, one where success in both business and sport can be perfectly balanced like a gymnast who has earned her gold medal.

9. Tomorrow you will play better than today, and better the next day than tomorrow. Learning new things, practicing new skills and rising above anything that has held you back as you leave the gate of comfort and enter a new experience of peak performance.

10. Go ahead, stand right here in this office, remaining in hypnosis, by adapting the perfect stance, or the perfect position and in your own mind create the motion you know creates excellence and as you stand here in the office, execute it through mime. It might feel silly at first to not hit a baseball for example, with a bat you are not holding – but the mind that creates is a mint creates reality, and by mentally rehearing and even doing it here until you are confident you have mastered the subtleties of success, you have retrained the brain, which moves the body, and

connects with the ball. And so by doing this here, the correct process is committed to the mind, to the body, and the next time you take to the field, the new reality will be exactly as you have planned it here.

11. Let go of your failures, right here and right now, recasting them not as failures but as learning experiences, taking what you need to know from every situation – no matter how disappointing – they might have been in the past to your new reality of having learned from each event and taking action not only to avoid a repeat, but to better your performance in every opportunity.

12. Breathe in now. Pay attention to the breath. Each breath marks each moment, and you have an ability to use the breath now, to mark the perfect moment to make your move.

Pain Control

1. Right now think of the word cool. Cool. Cool like the top of a refrigerator or cool like a fall breeze. When you think of this coolness can you create an awareness of coolness, even focusing on that place where pain had radiated and really fixating on that spot while thinking of cool? And notice when you do, just the thought of cool, calms the intensity of pain.

2. What color is your pain (await response). Now bring you attention to the center of your pain, but also imagine the pinpoint of a Chroma color wheel, in the center of that

pain. And imagine moving the wheel to change the color palate to a more conformable color. In your mind do that now – what color is more comfortable for you?

3. Have you ever been happy, but also sad? Have you ever been both excited by scared? I know that you feel pain, but at what level can you also create and be comfortable? Can you ow amplify that awareness of comfort, even if it's in a different part of the body? I bet you can even make the part of you that holds comfort as comfortable as you feel pain.

4. Notice your pain. Rather than wishing it away, or trying to make it stop, what would happen if you really observed it, studied it and allowed yourself to feel it. So often when we try to get rid of something it simply amplified awareness, it become like an itch we keep scratching to make it stop. And so rather than trying to stop it, focus on it, study it, follow it. Notice where the pain starts, where the pain stops ad where it is more intense and where it is less intense.

5. Imagine your hand is like a magnet, able to pull pain like a magnet can pull mettle scraps. I bet when you were a kid you played with a magnet and watched it attract to it metal shavings. And of course, you also learned that to create a magnet, you rubbed one magnet on another piece of metal. My hand is a pain magnet, and I'm placing it on top of your hand, and now your hand is a pain magnet. Go ahead, test it out, hold it over your pain, and notice how much of it is drawn out and removed by the

magnet. And now shake your hand, shaking loose the pain it has attracted, letting it simply fall to the ground and disappearing.

6.   Listen to your pain. Is it screaming, or yelling or making noises in your mind? If pain can come with sound, it can also come with a volume know, and imagine reaching out and turning the volume down. Maybe half the volume. Or maybe all the way to silent.

7.   Imagine pain is a target, and you can drive a bullet right through that target, blowing out the center of the target. At the moment you imagine the bullet passing through, what is the feeling in that spot? Feels pretty amazing doesn't it?

8.   Sooner or later everything changes, and you know even pain will disappear. Let go of it now, letting it leave early.

9.   A person may even give themselves permission to heal, to feel better and even to smile when things are hard or the body hurts.

10.   Imagine a single white puffy cloud floating across the sky and into the distance. As if moves towards the horizon, becoming smaller and smaller, cast all your pain into that cloud and let it be carried away . . . moving towards the horizon and disappearing . . . .

11.   By following the doctors suggestions for icing and healing your pain, you will notice that the results are exactly as promised.

12. Imagine a pain killer being injected right into the center of the pain. Notice what it feels like as the skin absorbs that pain killer, noting that there are no side effects, no prescriptions to go get, just an ability of the mind, to create the same awareness as if you have received an injection.

Suggestions Based On Erickson Language Patterns

Milton Erickson, M.D. was the psychiatrist who popularized medical hypnotherapy in the 1950's to the 1980's. His work is credited as being the foundation to many modern approaches to hypnotherapy. In particular, the language patterns he used, have been modeled to replicate his results with many clients with many presenting problems. The suggestions below are "formulas" based on his language patters. You can learn more about "The Milton Model" and hypnotic language patters in my book *"Speak Erickson Ian: Mastering the Methods of Milton Erickson."* is considered a classic in the arena of Ericksonian hypnosis by many teachers of hypnosis and you can get a copy from any bookseller or from www.SubliminalScience.com

1. <u>I wouldn't</u> tell you to (let go of your pain) and you can't disagree with me, <u>because</u> I said I wouldn't tell you.

   By structuring the suggestion this way, if forces the subconscious mind to process "what wouldn't be" told to the client. It avoid paternalistic suggestion and the linkage <u>because</u> is a linguistic maneuver that creates causation.

2. <u>Will you</u> (do this or that) <u>or</u> will you (this or that) <u>or</u> will you? Will you let go quickly, or will you take another session before you let go, or will you let go in your own time, regardless of our schedule?

   Structuring a suggestion in a way that uses <u>Will you</u> and <u>or</u> forces one of several choices to be made. These are often faked alternatives, and each choice often leads to the same outcome even if the process is different.

3. <u>A person may</u> easily find the answers <u>because</u> they have learned hypnosis

   <u>A person</u> is a generic term, and the subconscious mind will put itself in the category of a person, making the suggestion personal yet non-specific. <u>Because</u> explains why the first part of the suggestion is acted upon, creating congruency between the unconscious and conscious mind.

4. It is <u>as-if</u> change is easy and instantaneous

   <u>As-if</u> statements are presuppositions that are quite powerful. As you write your own suggestions, is <u>as-if</u> statements regularly.

5. <u>You might find</u> it is less difficult that ever imagined to (go into trance, quit smoking, hit a goal, etc.)

   <u>You might find</u> forces a person to look for something. And of course, we find what we seek!

6. <u>A person could (clients name)</u> quit smoking in just one session.

    By using a client's name in a session you instantly make the suggestions something for them, from them and give them ownership of the suggestions.

7. <u>Can you really enjoy</u> going into a deeper trance, and using this time to learn, empower and success?

    Asking questions, forces a client to look for answers. Structure almost any suggestion as a question, you can have confidence that the subconscious mind will discover the answer.

8. <u>It is easy</u> to go into hypnosis <u>now, isn't it?</u>

    Again, a question, coupled with an indirect suggestion. <u>It is easy</u> is a direct suggestion, <u>now</u> tells them when, and <u>isn't it</u> is a rhetorical question.

9. <u>You may not know</u> if you are reaching the deepest levels of trance, or can go deeper

    By stating they might not know something, causes a client to see what they do know.

10. <u>What happens when you</u> close your eyes and realize you have <u>now</u> stopped smoking?

    <u>Now</u> tells then when to experience or do something.

11. <u>Now</u> is the time.

12. <u>Try to resist</u> success and you will find that failure in your attempt to resist, leads to success.

<u>Try</u> is a word that implies failure. So it can be used to create paradoxical suggestions.